EMPIRE FORESTRY AND THE ORIGINS OF ENVIRONMENTALISM

What we now know as environmentalism began with a series of land reservations in 1855 in British India, spreading during the second half of the nineteenth century until over 10 percent of the land surface of the earth became protected as a public trust. Sprawling forest reservations, many of them larger than modern nations, became revenue-producing forests that protected the whole "household of nature," and Rudyard Kipling and Theodore Roosevelt were among those who celebrated a new class of government foresters as public heroes. These foresters warned of impending catastrophe, desertification, and global climate change if the reverse process of deforestation continued. The empire forestry movement spread through India, Africa, Australia, New Zealand, Canada, and then the United States to other parts of the globe, and Gregory Barton's pioneering study is amongst the first to look at this movement, and thus the origins of environmentalism, in global perspective.

Born under imperialism, environmentalism today is as profound a global movement as that for democracy itself. Ironically it is in those former colonies where environmentalism took shape that its future, and the future of nature, seems least assured. *Empire Forestry and the Origins of Environmentalism* is a major contribution to the understanding of what is perhaps one of the most powerful political and social forces of modern times.

Raised in Oregon, Gregory Barton is Professor of British, Colonial and Environmental History at the University of Redlands, California. He is the editor of *American Environmentalism*. This is his first book.

T0269377

Cambridge Studies in Historical Geography 34

Cambridge Studies in Historical Geography encourages exploration of the philosophies, methodologies and techniques of historical geography and publishes the results of new research within all branches of the subject. It endeavors to secure the marriage of traditional scholarship with innovative approaches to problems and to sources, aiming in this way to provide a focus for the discipline and to contribute towards its development. The series is an international forum for publication in historical geography which also promotes contact with workers in cognate disciplines.

For a full list of titles in the series, please see end of book.

EMPIRE FORESTRY
AND THE ORIGINS OF
ENVIRONMENTALISM

GREGORY ALLEN BARTON

CAMBRIDGE
UNIVERSITY PRESS

CAMBRIDGE UNIVERSITY PRESS
Cambridge, New York, Melbourne, Madrid, Cape Town, Singapore, São Paulo

Cambridge University Press
The Edinburgh Building, Cambridge CB2 8RU, UK

Published in the United States of America by Cambridge University Press, New York

www.cambridge.org
Information on this title: www.cambridge.org/9780521814171

First published 2002
This digitally printed version 2007

A catalogue record for this publication is available from the British Library

ISBN 978-0-521-81417-1 hardback
ISBN 978-0-521-03889-8 paperback

For my mother,
Ina Mae Russell Barton

Contents

Illustrations

Acknowledgments

Many people have helped me along the way to make this book possible. Richard Mills at the Oxford Forestry Institute guided me to much valuable material, particularly to early conservation photographs in the Plant Sciences Library collection. I am indebted to Gareth Griffiths and the archival volunteers at the new British Empire and Commonwealth Museum in Bristol, to the staff at the British Library, the Bibliothèque Nationale, the India Archives in New Delhi, the Library of Congress and the National Agricultural Library in Washington, DC. Generous funding from the Fulbright Commission enabled me to research in the subcontinent and to share my ideas at Dhaka University and the Asiatic Society of Bangladesh. The director, Sirajul Islam, gave me much helpful scholarly input. I also thank the director of the National Library of Bangladesh, Md. Shahabuddin Khan and the director of the National Archives of Bangladesh Mr. Chowdhury. I am also grateful to Shibbir Haq, a friend and colleague who provided much hospitality and care in Dhaka, and who guided me to many remote forest areas in northeast India, Nepal, Bangladesh, and Myanmar.

At Northwestern University, where this book began, I benefited from the help and advice of John Bushnell, Paul Friesema, T. W. Heyck, my dissertation advisor, and Harold Perkin. To Harold Perkin I owe a great debt and special thanks. From him I learned the love of writing "big picture" history and observed first hand the habits of mind that a great historian brings to his craft. He read many drafts of the manuscript, suggesting ideas and improvements throughout, and never ceased to make himself available at every stage from dissertation to book. To Michael Williams I also owe much. As the world's leading forest historian, he reviewed early drafts of articles that explored the themes of this work, and despite their many flaws recommended them for publication. He also selflessly aided my efforts to bring to publication an argument that he concurrently formed. I am grateful for the advice of Michael Heffernan, Alan Baker, Richard Grove, and the editors at Cambridge University Press. Matthew Carrell aided me with the conservation photographs in this book, providing much-needed technical expertise. Dr. Elias Stinson generously shared his brilliant ideas on nature, religion, and empire. Mary Curry, cited in this book as Mary Ledzion, became an invaluable friend who, born in India and raised with the Indian Forest Department, explained much that cannot be found in books. Ben Longrigg, also raised in the lap of the Indian Forest Department, generously provided travel funds to Britain and access to his father's control journal, along with his valuable advice and friendship. My mother laboriously read and corrected countless drafts in her forest cabin in Zigzag, Oregon, while my partner T. Neal patiently encouraged me as I traveled and wrote. I am grateful to them all. The faults of this book, are, needless to say, entirely my own.

1

Introduction

When and where did the environmental movement begin? Stepping back from the limitations of national history, this book examines the question of environmental origins on a global scale. In the late nineteenth and early twentieth century the most sweeping environmental initiatives emerged under the auspices of British imperialism. As the following study will show, hard-headed environmentalists and legislators found in empire forestry a ready-made model to persuade the public that the reservation of vast areas of the public domain would serve settlers, industrial development, governmental revenue, *and* environmental purposes. Empire forestry resolved the tension between romantic preservationist notions and *laissez-faire* policies. This book traces the international trail of environmentalism from India, under Lord Dalhousie's Forest Charter, to the British colonies in Africa and Australasia where it matured and, finally, to Canada, the United States, and other parts of the globe where environmentalism permanently entered the pantheon of political creeds.

By the First World War a large area of forested land around the globe lay in the public trust, managed by a professional cadre of government foresters. In the British colonies alone the crown had environmentally protected a land mass equal to ten times the size of Great Britain. Concurrently in the United States, after transferring 1 billion acres of public land into private hands in the early and mid 1800s (approximately one-half of the land mass of the continental United States) a change suddenly occurred. Congress authorized the president to set aside forest lands by proclamation and began America's process of environmental protection that would lead eventually to setting aside 15 percent of its land mass for various forms of protection and public use.

By 1928 British foresters managed environmentally every major forest type in the world. By 1936 the British Empire included a quarter of the land surface of the world, and of this, forests constituted one fourth. Fifty separate forest services protected not only trees but also soil, water, and – so foresters believed – the climate of entire continents and regions. Empire forestry triumphantly claimed credit for this achievement and served as an example for much of the reserved forest areas outside the British colonies. Out of a total empire of 9,737,660 square miles, Whitehall approved 2,465,530 miles as classified governmental forests, approximately 25 percent of British possessions and 8 percent of the land surface of the world.[1] Add to this the protected areas of the Republic of China and the United States, which consciously mimicked empire forestry, the figure rose to over 10 percent of the land area of the planet. Only the Neolithic and industrial revolutions compared to the impact of this third global revolution in land use.[2]

Imperial forestry experts promoted the very modern-sounding proposition that deforestation led to devastating changes in climate.[3] Forests in India, Australia, New Zealand, Canada, and the far-flung colonial empire in Africa, Latin America, and the West Indies constituted a global environmental laboratory with innovative strategies and new management techniques, watched attentively throughout the world. In India the forest department regulated a timber industry that employed over 5 million people, managing tropical rain forests, mangrove trees, tropical deciduous trees, dessert scrub, temperate broad-leaved woods, various conifer species, and European-like forests that grew on the foothills of the Himalayas. Burma in 1928 had fully one-half of its total area under forest canopy and profitably managed.[4]

In Canada the forest industry employed 244,000 people, the forests stretching in a great emerald belt from the seaboard of Quebec to the islands of British Columbia. In this colossal domain, Douglas fir led the volume of trees, followed by cedar and hemlock. South of Georgian Bay to eastern Quebec lay great reserves of maples, oaks, and hickories.[5] In Australia, though forests covered only 3.3 percent of the land area, empire foresters saved broad-leafed eucalyptus woods composed of several hundred species. In New Zealand the government preserved the kauri forests for soil, water, and climate protection, taxing the proceeds of timber sales to pay for the program. The government supplemented natural forests with commercial plantations, including indigenous conifers. In 1935 the New Zealand forest service managed over 12,000 square miles of state forests and valued a variety of ecological concerns equally with commercial timber extraction.[6]

In South Africa, with a small forested area of less than 4 percent of the land mass, the forest department reforested thousands of square miles with broad-leafed evergreens. In Southern Rhodesia the forest department governed 88,000 square miles of forest, much of it in private hands, and managed 6,000 squares miles of national parks, game reserves, and protected forests. In the colonial empire of the

[1] D. Brandis, and A. Smythies ed., *Report of the Proceedings of the Forest Conference held at Simla, October 1875* (Calcutta, 1876), 48. Opinion differed on whether forests affected the climate of whole continents or were a local effect only. Dietrich Brandis, Inspector General of Forests in India, took the latter view. Roy Robinson, "Forestry in the British Empire," *Journal of the Royal Society of Arts* 84 (1936): 795, 796.

[2] Robinson, "Forestry," 779. Harold Perkin, *The Third Revolution: Professional Elites in the Modern World* (London and New York, 1996), 2–4.

[3] Robinson, "Forestry," 779.   [4] ibid., 781.   [5] ibid., 782.   [6] ibid., 785–789.

1 Reserved teak forest in Burma. Empire foresters discovered that to preserve teak they had to preserve the whole "household of nature," including hundreds of other plant and tree species. 1914.

West Indies, West Africa, Kenya, and Ceylon, the Colonial Office administered vast areas. Mahogany, greenheart, pencil cedar, satinwood, and ebony supplied a thriving timber business, while gum trees in British Honduras, jelutong in Malaya, and cacao in the Gold Coast all required the maintenance of stable forest conditions for soil, stream flow, and humidity.[7] All this the imperial administrators managed at a profit by designing a "demonstrated use" area for industry and agriculture.[8] By 1928 fifty separate forest departments served the empire, with 1,500 officers, native junior officers in the tens of thousands, and 1,200,000 square miles of revenue-producing forests.[9]

Empire forestry here refers to forestry as practiced in the British colonies and, retrospectively, to forestry practiced from the inception of colonial conservation in 1855.[10] The term *empire forestry* developed at the apogee of British colonial forestry, and originated with the first British Empire forestry conference, which met on July 7, 1920 at the Guildhall, London. The forest services of India and Canada

[7] ibid., 789–793.
[8] Demonstrated use areas included access by the indigenous population for grazing and firewood.
[9] Robinson, "Forestry," 779.
[10] See the opening remarks of the Lord Mayor of London and Lord Lovat, *The British Empire Forestry Conference* (London, 1921), 1, 2.

initiated the meeting, requesting the British government to sponsor an all-empire forestry conference with delegates from all the forest services of the colonies (concurrent with the empire timber exhibition of 1920) to exchange ideas, coordinate policies, and collectively take stock of low timber supplies after the First World War. From this meeting, held once every four years, grew the Empire Forestry Association, the Imperial Forestry Institute, and the *Empire Forestry Journal.*

But how exactly, in an age of *laissez-faire*, did empire forestry arise? With an official policy of settlement and development, how did such vast areas of land come to be protected? What shift of attitude or belief divorced public opinion from *laissez-faire*? Why did the public embrace governmental intervention and environmentalism? How did public ownership of land come to be celebrated, with a new and barely defined professional corps of government foresters such as Dietrich Brandis and Gifford Pinchot feted as popular heroes?

Among scholars, environmental history has a historiographical canon largely rooted in the American scene. Why? Because environmental history, as distinct from the practice of conservation, essentially arose in the United States. From Frederick Jackson Turner's 1893 *Significance of the Frontier in American History* to Walter Prescott Webb's *The Great Plains* in 1931, and to James Malin's *The Grassland of North America* in 1947, a fascination for epic history of a peculiarly geographical nature developed, one that featured the dichotomy of a civilized people (Americans) invading pristine nature (the West), that pitted nascent consumer culture against raw and unexploited wilderness or, in the case of William Cronon's *Nature's Metropolis*, that pitted belching factories and towering skyscrapers over and against the good red earth, dominating, seducing and defining the Great West.[11]

Environmental scholars have also written an environmental history that was, as Donald Worster says, "born out of moral purpose." It drove the production of scholarship in the field.[12] Since the 1960s certain scholars insisted that environmental history, like a fugitive, was radical and subversive, riding on the horse of the Romantic movement, leaping to the new mount of literary nature-writing and then landing at last on the progressive saddle of Theodore Roosevelt.[13]

---

[11] Frederick Jackson Turner, "The Significance of the Frontier in American History," *Annual Report of the American Historical Association for 1893* (Washington, DC, 1894); W. Webb, *The Great Plains* (Boston, 1931); J. C. Malin, *The Grassland of North America: Prolegomena to its History* (Gloucester, MA, 1947). For an interesting analysis of the historiography of environmental history in the United States, see Michael Williams, "The Relations of Environmental History and Historical Geography," *Journal of Historical Geography* 20 (1994): 3–21. See also R. White, "American Environmental History: the Development of a New Historical Field," *Pacific Historical Review* 54 (1985): 297–335. William Cronon, *Nature's Metropolis: Chicago and the Great West* (New York, 1991).

[12] Donald Worster, *The Ends of Earth: Perspectives on Modern Environmental History* (New York, 1988), 290.

[13] Samuel Hays, *Conservation and the Gospel of Efficiency: the Progressive Conservation Movement 1890–1920* (New York, 1959); Richard Nash, *Wilderness and the American Mind* (New Haven, 1967). For a review of conservation literature see L. Rakestraw, "Conservation Historiography: an Assessment," *Pacific Historical Review* 41 (1972).

Marxist scholars add a new home for environmentalism – New Left history – where nature triumphantly shares a "bottom-up" land ethic and where biota and landscape assume moral, if not legal rights.[14]

Currently a problem in the field is that environmental historians find themselves unable to bridge the gap between the intellectual and cultural foundations of environmental inspiration on one hand and the first implementation of forest reservations on the other. As one environmental scholar reasoned, the nature of environmental history must be broader than American history alone and would have to incorporate institutional and economic realities in a global context, a "mutual," not an "unidirectional" history.[15] But a global history of environmentalism in the nineteenth and early twentieth century would engage an unpopular leviathan: empire. Because of the democratic aversion to empire, scholars have been hesitant to notice the imperial origins of environmentalism. Devotion to the idea that environmentalism is a "subversive science," sensitive to democratic aspirations, and opposed to an "imperial science," fed a reluctance to recognize empire as the proximate cause of environmental innovation.

Alice Ingerson, editor of *Forest and Conservation History*, laments how environmental historians often look at the nineteenth century as a "black box" from which to pull out explanations suitable to their point. All too often environmental ideas are summarized without reference to how Victorians related to their natural environment. The attempt by Thomas Lyon to characterize the whole of pre-1945 ecological history is typical: we are informed, unhelpfully, that "The roots of wilderness thought . . . go deep."[16] Likewise the back cover of David Evans' *A History of Nature Conservation in Britain* announces that the book is "the first complete history of the British nature conservation movement."[17] Yet Evans gives scant treatment of any development before 1890 and very little until the Second World War. Evans' "complete history" is not far off the mark because, unfortunately, environmental history of the nineteenth century is practically nonexistent.

Donald Worster, to his credit, initiated a few steps in the right direction. His book, *Nature's Economy*, makes a broad-stroke attempt to separate the moral and political motivations underlying environmental history and to combine scientific, romantic, and mystic notions of nature into an understandable history. However, he urges, along the lines of Theodore Adorno, the separation of "imperial" from "subversive" science in order to oppose transcendent morality and nature against the western (and heartless) empirical tradition. He also clearly separates a supposed democratic and reformist environmentalism from an empirical, capitalistic, and imperial science. This approach contains many flaws, particularly the uncritical

[14] See Richard Nash, *The Rights of Nature: a History of Environmental Ethics* (Madison, 1989).

[15] L. J. Bilsky, ed., *Historical Ecology: Essays on Environment and Social Change* (Port Washington, NY, 1980), 4, 8.

[16] Thomas Lyon, review of *The Idea of Wilderness: from Prehistory to the Age of Ecology* by Max Oeschlaeger, *Forest and Conservation History* 36 (1992): 146.

[17] David Evans, *A History of Nature Conservation in Britain* (London, 1992).

acceptance of Adorno's thesis that reads into the past an alliance of progressivism and environmentalism. But Worster deserves credit for opening environmental history into a European-wide perspective that incorporates a multilayered inquiry into the history of science, botany, economics, and imperialism.[18]

Imperialism and environmentalism have a shared past that scholars cannot gloss over. Those who think imperialism and western expansion a "good thing," as do conservatives touting parliamentary democracy and western standards of human-ism, and those who think imperialism a "bad thing," as do many third world historians and postmodernists, may here agree that environmentalism is police action, inseparable from western conceptions and attitudes. The conservative and the anticolonialist postmodernist may begin with ideological cleavage, like the dual tips of a horseshoe, and yet agree at the arch. The more conventional esta-blished left, who assert a radical cleavage between the agent (imperialism) and the effect (environmentalism) offer an unhistorical past. Surely the antiscience science (as Worster dubs ecology) may not function as a challenge to imperialism and established "European male hegemony," as many environmental historians from Worster, Carolyn Merchant, to Richard Grove suppose.[19]

In *Green Imperialism*, Richard Grove also struggles to explain the origins of environmentalism. He rightly identifies environmental ideas emerging from a co-terie of professional scientists on the colonial "periphery" of the European empires. Following the utopian, physiocratic, and medical contributions to the history of climate theory, he especially examines the effect of deforestation on Mauritius and St. Helena. His work examines the writing of selected surgeons to indicate that tropical deforestation by 1850 was conceived of as a global problem.

Grove also asserts that climate theory came to the attention of the British through the experience of tropical islands. But while it is true that the British observed climate change on St. Helena and Mauritius, the theory of climate change by deforestation filtered through a very wide variety of sources. Grove particularly misses how observers analyzed the effect of deforestation in the United States and Europe, downplaying in particular the significance of G. P. Marsh's *Man and Nature*, a classic text of environmental theory in the nineteenth century. Grove's dense work all too often merely traces historical instances that appear to resemble "environmentalism." Oddly, environmentalism is never defined. Those nineteenth-century journals that discuss climate theory he leaves largely unexplored. Grove stops the investigation precisely at the point where climate theory began to affect legislation.[20]

But Grove would agree with Thomas Richards in *Imperial Archive* that the drive for knowledge and the penchant to divide and catalogue the world – central

---

[18] See Donald Worster, *Nature's Economy: a History of Ecological Ideas* (Cambridge, 1995).

[19] Worster and Merchant do not make this argument alone. See Paul Sears' "Ecology – a Subversive Subject," *BioScience* (July 1964): 11–13.

[20] See Richard Grove, *Green Imperialism: Colonial Expansion, Tropical Island Edens and the Origins of Environmentalism, 1600–1860* (Cambridge, 1995), 1.

to the environmental project – grew from taking stock of the "inventory" of imperialism.[21] Part of that inventory is the discovery, subjection, demarcation, and effective management of nature. Indian imperial officials inaugurated a modern forestry management system that spread from India to much of the world. This was most likely to happen, not under a democracy where a majority of users – peasants, craftsmen, and traders – could block reform, but under an authoritarian regime where colonial overlords could, for better or worse, impose their control. These environmental innovations, born of empire, mark the first clear boundaries of environmentalism.

This book follows a chronological and geographical order, tracing the birth of environmental practice in British India and following its spread to Australia, New Zealand, Africa, Canada, and then outside the empire to the United States. As Stephen Pyne notes in *Burning Bush: a Fire History of Australia*, most forest department officers in British employ passed Greater India.[22] These same foresters retired to teach at forestry schools in Britain, Europe, the colonies, and the United States – most notably at Cooper's Hill (in Britain), the University of Edinburgh, Cambridge, Oxford, Aberdeen, Bangor, Yale, Sydney, and Toakai (in Cape Town).[23] Read, discussed, and imitated, they not only taught new generations of foresters but commanded the attention of popular magazines and scientific journals.

The years 1855 to 1945 span the birth of Lord Dalhousie's Forest Charter to the end of the Second World War. Most scholars agree that the Second World War marks the shift from a conservationist environmentalism to a modern environmentalism focused on pollution, health, and work safety. One source during the 100-year period from 1855 to 1945 has been indispensable. The *Indian Forester* chronicled the growth of empire forestry and its practice, and has been neglected in environmental history. The first India-wide forestry conference in 1872 commissioned, under the supervision of Baden Powell, a "Forest Magazine" that collected and commissioned information on forestry that could be made "generally interesting . . . not confining it . . . to technical forest matters" only, and published quarterly. This journal, along with government bulletins, special reports, and parliamentary inquiries form the base of my sources.[24]

---

[21] Thomas Richards, *The Imperial Archive: Knowledge and the Fantasy of Empire* (London, 1993). To understand the relation between the archive and imperialism, see Richard Garnett, "The British Museum Catalogue as the Basis of a Universal Catalogue," *Essays in Librarianship and Bibliography* (London, 1899). A history of the British Museum and its role in politics can be found in Barbara McCrimmon's *Power Politics and Print: the Publication of the British Museum Catalogue, 1881–1900* (Hamden, CN, 1981).

[22] Stephen J. Pyne, *Burning Bush: a Fire History of Australia* (New York, 1991), 260.

[23] *British Empire Forestry Conference* (London, 1921), 23; "The Training of Candidates and Probationers for Appointment as Forest Officers in the Government Service." Report of a committee appointed by the Secretary of State for the Colonies, July 1931, Colonial no. 61, 1931; E. B. Worthington, *Sciences in Africa* (Oxford, 1938), 17.

[24] B. H. Baden Powell and J. C. Macdonell, *Report of the Proceedings of a Conference of Forest Officers Held at Lahore, January 2 and 3, 1872* (Lahore, 1872), 84–85.

Berthold Ribbentrop and E. P. Stebbing, onetime head of the Indian Forest Department, wrote irreplaceable histories of forestry in India. Both authors cobbled together a series of governmental reports to form books. Both gave detailed observations in a province-by-province narrative that left the reader to imagine an India-wide chronological narrative. The use of their work represents a radical reorganization of their information, in order to piece together an India-wide treatment that is both chronological and issue oriented. Also essential have been the reports of D. E. Hutchins, who as a forester-at-large roamed Australia, Cyprus, Africa, and New Zealand and wrote reports for various governmental bodies. Franklin Hough, America's first forestry agent, Charles Sargent, botanist and popularizer of forestry practice, and Gifford Pinchot, who oversaw the birth of the forestry department in the United States, are the primary sources for the chapter on American environmentalism.

2

The great interference

Environmentalist thought before the 1960s revolved around forests and their preservation. For instance, it was only in the 1980s that the journal *Forest and Conservation History* (founded 1957) began broadening the concept of environmental history beyond forest issues alone; the publication is now called the *Journal of Environmental History*.[1] Early advocacy for preservation focused on forest land for a number of reasons. Timber supply and revenue questions always demanded the attention of governments. But climate theories that explained how forest lands affected rainfall, along with soil preservation, water flow, animal life, and the preservation of a variety of forest flora and fauna made forestry the most pressing environmental issue of the late nineteenth and early twentieth centuries.

Empire foresters usually understood the broader implications of their work, and the effect of forestry practice on the environment. In 1872 Baden Powell admonished his officers at the first India-wide forestry conference to "regard the planting and restoration of our divisions as your chief business." "Never," he instructed "consent to work as if the felling of timber was the great work of life, and as if the provision of a few rupees in the Budget under the planting head . . . was all that is needed by way of supplement."[2]

Defined broadly, environmentalism means merely the advocacy of a proper balance between humans and the natural world. Certainly a history of modern environmentalism is a history of the relationship of people with their environment, particularly the history of advocacy and preservation.[3] More specifically, Worster defines environmentalism as

---

[1] Michael Williams, "The Relations of Environmental History and Historical Geography," *Journal of Historical Geography*, 20 (1994): 3.

[2] Baden Powell and J. C. Macdonell, *Report of the Proceedings of a Conference of Forest Officers Held at Lahore, January 2 and 3, 1872* (Lahore, 1872), 90.

[3] *Merriam-Webster's Collegiate Dictionary* (New York, 1979) defines environmentalism as, "2. Advocacy of the preservation or importance of the natural environment; especially the movement to control pollution." Pollution control, though not unknown to nineteenth-century environmentalists, defines contemporary environmentalism rather than nineteenth-century environmentalists who, though not unaware of pollution of water and air, focused on the more pressing need of preservation.

a set of environmental ideals demanded by an urban, industrial society. The period from 1860–1915 saw the emergence of these ideals, a body of thought that we can call environmentalism. That man's welfare depends crucially on his physical surroundings was a central premise of the new environmentalism. Another sacred assumption was that it is better for society, through the agency of experts, to design and direct the development of the landscape rather than leave the process in the hands of untrained, self-interested men. Coordinated public planning would end what was viewed as the haphazard and exploitive practices common in the *laissez-faire* approach. A third dictum of the emerging environmentalism, and perhaps the most important, was the belief that science and scientific methods must become the chief foundation on which environmental plans would be built.[4]

Worster equates the conservation movement with environmentalism, asserting as most environmental scholars do that the conservation of forest lands constituted an early phase of environmentalism – even when those lands were set aside only for issues of timber supply and revenue. Only after the Second World War did the focus of the environmental movement shift to pollution and health concerns.[5] Forest history is the history of how humans have related to much of the natural world, and therefore it has played a central role in the history of environmentalism.

The forest remained the primary focus of environmental concern before the Second World War. Forest clearance in northwestern Europe caused widespread concern over fuel supply, timber, climate, and water flow. In the Middle Ages and early modern period northwestern Europe relied more heavily upon wood products than did the Mediterranean region. The invention of the stiff-collar yoke in the tenth century and the accompanying increase of harnessing power (the old cloth and leather yokes had wrapped around the neck of the animal and constricted its windpipe) radically increased the extraction of forest products.

In *L'attelage, le cheval de selle à travers les âges* (1931) Lefebvre des Noëttess argued that the transition to the new yokes took place under the Capetians, and had a direct bearing on man's increased ability to change the natural landscape. Also in this period, northwestern Europeans utilized watermills. While in Italy, Spain, and Greece the lower water levels often meant dry summer seasons that limited the productive powers of mills, in England, France, and Germany the higher water flow guaranteed a year-round supply of energy. The new hydraulic saw and the formula of abundant forests, draft power, and watermills added up to an increased ability to clear forests and utilize timber.[6]

---

[4] Donald Worster, *American Environmentalism: the Formative Period, 1860–1915* (London, 1973), 2.

[5] See ibid., 85–95; and Robert Gottlieb, *Forcing the Spring: the Transformation of the American Environmental Movement* (Washington, DC, 1993), 8.

[6] Bertrand Gille, "Le moulin à eau," *Techniques et Civilisations* 3 (1954): 1–12. For an introduction to how the English modified the land, see W. G. Hoskins, *The Making of the English Landscape* (Harmondsworth, 1955); W. G. Hoskins and H. R. Finberg, *Common Lands of England and Wales* (London, 1963); H. C. Darby, *An Historical Geography of England Before AD 1800* (Cambridge, 1951); Oliver Rackham, *The History of the Countryside* (London, 1986); Ann Bermingham, *Landscape and Ideology: the English Rustic Tradition, 1740–1860* (Berkeley, 1986). For a Europe-wide context see William A. Watts, "Europe," and Karl-Enst Behre, "The Role of Man in European

A new sensibility arose parallel to these technological innovations. Keith Thomas characterizes a dilemma, in the early modern period, between man's material needs on the one hand and a growing fondness for nonurban areas on the other. Facilitated in part by a closer study of animals and an erosion in the certainty of human uniqueness, this new sensibility "generated feelings that would make it increasingly hard for men to come to terms with the uncompromising methods by which the dominance of their species had been secured." While the material needs of society demanded raw material, the "ruthless exploitation" of animal life and forest areas met a certain amount of resistance.[7]

The earliest regulation of forest use occurred within the framework of custom and usage. Legal strictures tended to preserve traditional forest usage for every stratum of society – for king, church, nobleman, and peasant. Though deforestation sometimes raised concern, this concern did not amount to a modern conception of environmentalism, with all its varied implications of ecological balance, biota preservation, water flow, soil, air, and climate stability. As Clarence Glacken pointed out,

> The reservation of choice forest areas for hunting grounds is a type of forest use that has been mentioned frequently in modern histories of forests. The argument is that this was unwitting conservation, not because the need for such conservation was understood, but because royal or noble enthusiasm for hunting enforced exclusions of destructive intruders. In France, these policies engendered hatreds against the forests and their royal and noble owners that reached a climax in the French Revolution. Although there is no doubt that forest landscapes were maintained which would otherwise have been destroyed, this emphasis does less than justice to the history of forest use and to the complexities in customary practice and usage, especially in the later Middle Ages.[8]

It is true that English nobles in the thirteenth century gained from King John the Magna Carta and with it a forest charter.[9] But forest laws often forbade the creation of new forests in order to preserve arable land. Forest rules were not intended to affect "nature" and man's relation to it in any broad sense, but remained local and wholly pragmatic.[10] The rights of usage tended to narrow throughout the Middle Ages and early modern period to more precisely define local rights and exigencies.[11]

Vegetation History," in B. Huntley and T. Webb III, eds., *Vegetation History* (London, 1988). Richard Lefebvre des Noëttess, *L'attelage, le cheval de selle à travers les âges. Contribution à l'histoire de l'esclavage*, vol. I (Paris, 1931), 2–5.

[7] Keith Thomas, *Man and the Natural World: a History of the Modern Sensibility* (New York, 1983), 302–303.
[8] Clarence Glacken, *Traces on the Rhodian Shore* (Berkeley, 1967), 326.
[9] Doris Stenton, *English Society in the Early Middle Ages* (Harmondsworth, 1951), 98–118.
[10] Richard Grove makes this mistake in *Green Imperialism*. Earlier articles by Grove also explore the theme: see "The Origins of Environmentalism," *Nature* 3 (May 1990): 11–15, and, "The Origins of Western Environmentalism," *Scientific American* 267 (1992): 42–28.
[11] Roger Grand and Raymond Delatouche, *L'agriculture au moyen âge. de la fin de l'empire Romain au 16e siècle* (Paris, 1950), 430–432.

Two founding documents indicate a new reticence about man's modification of nature: John Evelyn's *Silva* and the French Forest Ordinance 1669. Under the English king, Charles II and following the devastation of the civil war, the Royal Society deputized Evelyn to publish an essay regarding the timber supply. The society feared that glass and iron factories unduly depleted England of wood, and that after a harvest of trees few members of the landed elite attempted to reforest their estates. Evelyn, drawing on quotations from the classics, argued in a spirited report to the society that a national forestry policy would resolve the tensions between forests, agriculture, grazing, and industry.[12] He suggested that the practice of forestry aided the defense of the realm by provisioning the navy and the economy. England needed oak for the ships, cider for the sailors, and charcoal as fuel for the glassworks and iron furnaces. Of the latter he complained that "Truly, the waste and destruction of our woods has been so universal, that I conceive nothing less than a universal plantation of all sorts of trees will supply, and will encounter [*sic*] the defect."[13]

The "devouring iron-mills" were ruining England, Evelyn argued, and the mills ought to relocate to New England, where a limitless supply of wood grew. Interestingly enough, however, he linked deforestation with climate by observing how deeply forested areas produced moist unhealthy air. Though he argued that landowners ought to replant trees for both aesthetic and economic reasons, he also concurred with the popular understanding that a forest erodes good health and "would hinder the necessary evolution (that is, the action of flying out or away) of . . . superfluous moisture and intercourse of the air." Thus to clear forests had the healthful effect of "letting in the air and the sun and making the earth fit for tillage and pasture" – forest land, "those gloomy tracts," have in England "now become healthy and habitable" because they have been greatly reduced.[14]

Another early forest act, the French Forest Ordinance 1669, passed under Louis XIV, turned a bewildering array of local customs and rights into a national and consistent set of regulations. The bill of 1669 reformed the general forest code of Melun, passed under Charles V in 1376. Colbert, minister for Louis XIV, feared that "France will perish for lack of woods" if additional action did not ensue; hence he appointed in 1662 a commission to investigate the abuse of forest lands

---

[12] Because of Bacon's assertive view of man as shepherd, steward, and scientist, he has never been included in the pantheon of Arcadian forerunners of preservation. This has obscured his role as inspiration for writers not only like Evelyn, but for an army of civil scientists and environmentalists who instigated environmental protection as an "enlargement of the bounds of Human Empire." Once the arcadian–imperial dichotomy is removed and environmentalism is seen as an integral part of science, he is no longer a foil but a precursor of environmentalism. See Francis Bacon, "The Great Instauration," in *The Works of Francis Bacon*, ed. James Spedding, vol. I (New York, 1872–1878), 39.

[13] John Evelyn, *Silva, or a Discourse of Forest-Trees and the Propagation of Timber in His Majesty's Dominions* (London, 1706), "To the Reader," no pagination.

[14] ibid., 30–34.

and the supply of timber for the navy. Colbert then divided France into eighteen *arrondissements* (*grandes-Maîtrises des eaux et forêts*). With special regulations for the fir forests of mountain regions (which supplied ships' masts), his ministry reserved one-fourth of the public forest area – including the forests of royal and corporate estates.[15]

Clearly both Evelyn's *Silva* and the French Forest Ordinance of 1669 had environmental implications. Both recommended replacing local regulation and customary usage with a coherent national policy. Both affected private owners. But in France, due to greater royal and corporate ownership and a more centralized bureaucracy, Colbert succeeded in applying the rudiments of a national forest policy. In England a landed elite held land with highly developed property rights that blocked the implementation of Evelyn's suggestions and retarded the evolution of a coherent national policy.[16]

Scientific forestry owes its origin to eighteenth-century Germany. Here "theories, practices, and institutional models" provided the "starting point" for national efforts in forestry. This tradition arose out of the cameral sciences, a system of science applied to governmental offices attempting to devise a profit to the state through methods of strict quantification and regulation. Forest management was one aspect of cameral science.[17]

In France, half of the forests had been at one time in the domain of a royal house. After the forest ordinance of 1669, other lands were soon managed under its authority. Some came into the hands of the state through the revolution of 1790, when the state confiscated ecclesiastical property and seized other properties from nobles. This gave the government of France in the nineteenth century a large base of land to manage, compared to the British home government or to the decentralized states of pre-Bismarck Germany. But the forests melted away with the growth of population and overuse. In 1791 there were 18,166 square miles of forests in France, a figure reduced to 3,792 miles of forests by 1876. The sale of land or

---

[15] The best history of the 1699 ordinance is still G. Huffel, *Economie forestière*, 3 vols. (Paris, 1904–1907). See also L. F. A. Maury, *Les forêts de la Gaule et de l'ancienne France* (Paris, 1867). A brief account is found in L. F. Tessier, "L'idée forestière dans l'histoire," *Revue des Eaux et Forêts* (January–February 1905); F. Bailey, "Forestry in France," *Indian Forester* 15 (1886). The text of the 1699 ordinance is found in *Jurisprudence générale. Répertoire methodique et alphabetique de legislation de doctrine et de jurisprudence*, ed. D. Dalloz and Armand Dalloz (Paris, 1849), 15. For an English translation, and other obscure sources on French forestry, see John Croumbie Brown, *French Forest Ordinance of 1669 with Historical Sketch of Previous Treatment of Forests in France* (Edinburgh, 1883).

[16] For a general history of forestry in England consider Charles Cox, *The Royal Forests of England* (London, 1905); N. D. G. James, *A History of English Forestry* (London, 1981); Mark Anderson, *A History of Scottish Forestry*, 2 vols. (Nelson, 1967). For individual forest histories see Oliver Rackham, *Hayley Wood: its History and Ecology* (Cambridge, 1975) and Colin R. Tubbs, *The New Forest: an Ecological History* (Newton Abbot, 1968).

[17] Henry Lowod, "The Calculating Forester: Quantification, Cameral Science, and the Emergence of Scientific Forestry Management in Germany," in *The Quantifying Spirit in the Eighteenth Century*, ed. Tore Irängsmyr, J. L. Heilbron, and Robin E. Rider (Berkeley, 1990), 315–317.

timber by the government accounted for most of the reduction. The sale of forest land in France did not stop until after 1870.[18]

French forestry practice included the definition of the rights of the user, building roads and structures, replanting, financial planning, grazing, hunting, and the administration of penalties for offenses. Aside from forming a highly organized model of administration and training, French foresters contributed successful schemes for afforestation. France suffered heavily from the effects of grazing, with forests in sensitive areas dying off and leaving the soil vulnerable to erosion. Solving this problem meant replanting forest land to benefit farmland and reduce flooding.

The grazing of sheep and goats caused severe damage in mountain regions of France, particularly in the southern Alps and the immediate lowlands. With deforestation, flooding worsened, debris blocked rivers, and villages, bridges, roads, and railways were often swept away. Foresters planted in deforested areas, particularly in the mountains, and helped create a legal structure to support the work of the state, forcing private landholders to replant. If the state owned the land, the forest department commenced the replanting and paid the cost; if private parties owned the land, the case would be judged individually, sometimes requiring the proprietor to replant without aid, sometimes requiring the proprietor to replant with the state defraying the cost.[19]

The French were also known in the late nineteenth century for work on the dunes of the west coast. Here winds blow sand continuously onto shore, which forms dunes. These dunes push inward and cover arable land, encroaching on the villages that cluster along the coastline. To fight this, since 1789 the French, under the public works department, and under the forest department in 1862, built wooden palisades and planted maritime pine with broom, gorse, and grasses. Frenchmen tapped the wood for resin, which paid much of the cost of planting and maintenance. Private landowners did the same, or the state planted at state expense, reserving the right to harvest the pine. In this way a complex maze of private and public lands have been managed and much coast land protected.[20]

Forest education was greatly admired outside France, in particular the school established at Nancy. This school, founded in 1824, trained all grades of forest officers. Admission was by open competition. In the late nineteenth and early twentieth century the program ran along these lines. Every year for two years students studied six and a half months of theory, two and a half months of practical instruction, and took examinations during the final month. The first year focused on sylviculture, the second year on working plans and forest management. Interestingly, the school included a professor of the German language among the many positions.

---

[18] United States, Department of State, *Forestry in Europe: Reports from the Consuls of the United States* (Washington, DC, 1887), 280.

[19] *ibid.*, 296–297.

[20] ibid., 298–299; See also Theodore S. Woolsey, Jr., *Studies in French Forestry* (New York, 1920).

It did this because of the heavy German influence on French forestry and the cross-fertilization between the two nations in sylviculture.[21]

Captain Ian Campbell-Walker, a prominent forester in the Madras Presidency in India, held German forestry in high regard. In a report of 1873 on forestry in Germany, Austria, and Great Britain, he admired the scientific and professional spirit of the German foresters. Britain was ahead in agriculture, but behind the Germans in forestry, he argued. There were reasons for this. Large state forests did not exist in Britain because private landholders owned most of the land. Only in India did the crown hold vast areas that could be converted into national forests. What then can be learned from the Germans? (1) A system of forest management; (2) survey techniques; (3) proper methods of taking inventory; (4) how to settle rights; (5) how to conduct experiments on measuring the rate of tree growth; and (6) how to suit the soil to the tree. This systematized scientific and carefully measured approach of German forestry gave Indian foresters, he suggested, a knowledgeable base for the scientific management of forests in India.[22]

Overly optimistic, he wrote his report in 1873, when Indian forestry had just begun to formulate its basic system of management. As Campbell-Walker, Brandis, and others were to discover, the differences in the sheer size of the land blocks, variations in climate, and the process of setting aside new land areas that did not have a clear tradition of rights, made the challenges of Indian forestry unique. The difference between managing uncontested forest areas and reserving new forest areas as a multiuse forest for the public good made the crucial difference between traditional forestry and the new conservation movement.

Campbell-Walker advocated the reservation of forest areas in the empire knowing full well the radical nature of the procedure. This differed markedly from the German approach. For instance, he advocated for the Madras Presidency, "the speedy definition, registration, and, wherever possible, buying up or commutation of all such rights." Since the Indians expected strong centralized leadership from an imperial government, he argued that the British were uniquely situated to act. He wrote: "I think an act and the appointment of a joint commission are necessary to enable us to do it."[23] He also acknowledged the role of propaganda in gaining public support for such a radical program, along with the liberal use of money to purchase forest rights outright or barter them in return for firewood privileges. Nothing in Germany or France compared to this radical program.

Even as Campbell-Walker began to delineate differences with German forestry, he made his admiration clear. An *Oberförster* could "tell the name, local and botanical of every tree, shrub, and plant," he wrote. He could "classify it, and state its uses" and also do the same for insects. The German foresters also knew the yield, rate of growth and revenue of each block, could estimate exactly the

---

[21] US Dept. of State, *Forestry in Europe*, 304.

[22] Ian Campbell-Walker, *Reports on Forest Management in Germany, Austria, and Great Britain* (London, 1873), 57–59.

[23] ibid., 59.

timber to be cut, where he would sell it, at what price, from what part of the forests it would come, how much would be replanted, how much left to natural regeneration, and what could be thinned without damaging the whole. The new conservation movement in the British Empire had much to learn and, later, much to teach European forestry.[24]

From teleology to progress

In the eighteenth century Enlightenment thinkers introduced new ideas about nature. In addition to the long-standing observation that climate affected humans and culture, certain thinkers argued for the first time that human activity affected climate. Alexander von Humboldt saw man's influence on nature as global, not just local or national. Humboldt saw wide global change – humans disseminating plants throughout the world, changing and modifying nature, pursuing agriculture, gathering plants in exotic climates, pressing nature into narrow channels, and leaving the scenery of developed countries with a larger population and a monotonous landscape.[25] A contemporary of Humboldt, Johann Herder, saw man "as a band of bold though diminutive giants, gradually descending from the mountains, to subjugate the earth and change climates with their feeble arms." Only the future would tell how far these diminutive creatures were capable of changing the earth.[26]

But the new understanding that humans changed the environment and even affected local and global climate did not necessarily foreshadow environmentalism. Charles Montesquieu, for example, who along with Humboldt and Herder stressed the interplay between the environment and human culture, also expressed the fear that underpopulation would allow nature to grow and prosper at the expense of civilization.[27] Hume and Kant, fully aware that human intervention affected climate, nonetheless applauded the healthful result of draining marshes and clearing forests. Forest clearance and cultivation, they believed, improved the New World – America in particular – with milder winters, cooler summers, and a general climatic improvement.[28]

---

[24] ibid., 62. For an excellent account of German forestry up to 1938 see Franz Heske, *German Forestry* (New Haven, 1938).

[25] Alexander von Humboldt, *Essai politique sur le royaume de la nouvelle-Espagne* (Paris, 1811), 25–30. For his definitive work on plant geography and nature as a social entity see *Edeen zu einer Geographie der Pflanzen nebst einem Naturgemälde der Tropenländer* (Tübingen, 1807).

[26] Johann Herder, *Outlines of a Philosophy of the History of Man*, trans. T. Churchill (London, 1800), 59.

[27] Felix Mourisson, *Philosophies de la nature: Bacon, Boyle, Toland, Buffon* (Paris, 1887). In his *Pensées*, Montesquieu had argued that the fish of the oceans were inexhaustible, that coal pits could happily replace forests, and that more exploitation was needed. Charles Montesquieu, *Pensées et fragments inédits de Montesquieu*, ed. Gaston de Montesquieu, vol. I (Bordeaux, 1899–1901), 179–182.

[28] David Hume, 'Of the Populousness of Ancient Nations,' in T. H. Green and T. H. Grose, eds., *Essays Moral, Political and Literary* (London, 1882), 356–360.

The Comte de Buffon hated regions never inhabited by humans, with gloomy forests that threatened safety and choked out the soft delights of a cultivated landscape. He saw uninhabited nature as a waste, and if not dangerous, then unhealthy. He saw man as noble, with nature his mistress. As a god he controlled torrents, conquered the tumultuous sea and turned nature's waste into fertile land. "How beautiful it is, this cultivated Nature! That by man's care it is brilliantly and pompously subdued!" (Qu'elle est belle, cette Nature cultivée! que par les soins de l'homme elle est brillante et pompeusement parée!).[29] Buffon even advocated cutting forests to increase the ground heat of northern Europe and thus create a more temperate climate. The vision to preserve or the boldness to restore nature to a pristine condition cannot be found in Buffon, Montesquieu, Herder, or Humboldt.

The idea of an orderly and designed world meant for many a nature infused with purpose and direction, while in other writers it held no purpose or direction at all. Hume and Kant, for instance, questioned the objective existence of nature altogether. Hume attempted to cut the biblical umbilical cord of nature that gave humankind a role of domination and stewardship.[30] By arguing against natural theology and its doctrine of providential design, Kant asserted that causality cannot be proven and that the very concept of nature is a construct of the human mind – an idea – with human intelligence attempting to set value and interpretation artificially to nature.[31]

Paradoxically, even the ontological doubts raised by Kant and Hume have not, as Glacken pointed out, eliminated teleology from the modern understanding of nature, but rather shifted the emphasis to "the idea of progress." This last notion did not necessarily deny creation or the idea of a designed earth, but did emphasize an unbounded optimism that humankind can and ought to manage his environment so that he materially improves his happiness. Certainly forest clearance could contribute to human control and happiness. But the idea of progress led to a broad-based management of nature that the ancients could never have foreseen. Humans could for the first time imagine utilizing vast fields of knowledge, observation,

---

[29] Comte de Buffon, "De la nature. Premiére vue," *Histoire naturelle, générale et particulière* vol. XII (Paris, 1749–1804), xii-xv.

[30] David Hume: "And what say you to the discoveries in anatomy, chemistry, botany? . . . These surely are no objections, replied Cleanthes: they only discover new instances of art and contrivance. It is still the image of mind reflected on us from innumerable objects. The discoveries by microscopes, as they open a new universe in miniature, are still objections, according to you; arguments, according to me. The farther we push our researches of this kind, we are still led to infer the universal causes of all to be vastly different from mankind, or from any object of human experience and observation." See *Hume Selections*, ed. Charles W. Hendel, Jr. (New York, 1927), 328–329.

[31] Immanuel Kant: "Nature . . . organizes itself, and does so in each species of its organized products – following a certain pattern, certainly, as to general features, but nonetheless admitting deviations calculated to secure self-preservation under particular circumstance . . . Therefore the organization of nature has nothing analogous to any causality known to us." See *Critique of Pure Reason*, trans. F. Max Müller (New York, 1902), 23.

and experience to effect a utilitarian stewardship over nature, while preserving and managing nature for its own sake and for future generations.

In the early 1920s J. B. Bury traced the idea of progress from Enlightenment optimism – a belief in the continuous progress of knowledge, which in turn led to greater control over the environment. Since the doctrine of progress "is a theory which involves a synthesis of the past and a prophecy of the future" it required an environmentalism of necessity – a rational belief that the earth would be habitable in the year 2100 AD or beyond.[32] This prophetic concern for the future of the earth and of society had its origin in the victory, in the seventeenth and eighteenth centuries, of the "moderns" over the "ancients." Though widely dismissed today as a poetical dialectic, for over a century the argument staked the central idea of the Enlightenment against a stubborn opponent, conservatism. If the world degenerated, as the advocates of the ancients contended, then the best poetry, philosophy, science, and political systems were in the past; minds could not be produced of equal mental vigor; grass is less green, the sun less warm, the prospects of the future bleak. An assiduous study of the ancients, conservatives argued, could at least preserve the best that had been produced for a declining culture and world.

But if the moderns prevailed, then the present age could be considered a high point of human history. Nature, according to the moderns, was still vigorous and would be so forever. Science would only improve the lot of humankind, and history could be seen as the life of a man who matured from youth to middle age, with no old age occurring. Why? Because the human race would always learn from the past and produce offspring as good or better than the world had seen. The world would become, increasingly, a handmaiden, subservient to the will of humans, made to be gentle and obedient, producing bountiful harvests, timber, minerals, and coal. The ocean would be a highway for commerce and equality between nations and people. Nature would be a genie loosed from the bottle of superstition to serve humans. But who would loose the genie? Science seemed the obvious candidate, with the Royal Society founded in 1660 and the Academy of Sciences founded in 1665. Both seemed destined to play just such a role. The Poet Laureate John Dryden optimistically serenaded the new societies who saw "remotest regions ... allied ... one city of the universe, where some may gain and all may be supplied."[33]

What did the moderns think of the past? As the Marquis de Condorcet argued, the past served as a bad example – we learn from the ancients by studying the mistakes of the past and by properly applying the physical sciences to the needs of men. This obsession with the future, regard for the mistakes of the past, respect for the physical sciences, and unbounded optimism of the capabilities of man, provided, as will be shown, the raw material out of which modern environmentalism had its origins.[34]

---

[32] J. B. Bury, *The Idea of Progress* (London, 1920), 5.   [33] ibid., 97.

[34] Interestingly, though Karl Marx and Charles Darwin have been used to challenge a teleological view of nature, both assumed that nature progressed and that nature had a purpose and an end. This sense of purpose is teleological. See Glacken, *Traces*, 550.

Imperialism

Nature, imperialism, science, and environmentalism are inextricably linked. No neat mental categories can be constructed with any justice to the historical record that entirely separates them. Europeans responded to the discovery and cataloguing of nature by an invigorated interest in the development of the natural sciences. The new maritime trade routes to Africa, India, the Atlantic and the Far East kindled this interest. National boundaries did not stop an English idea from circulating in Italy or Russia. Amateur scientists collected specimens on an international scale, with a European aristocratic (and rapidly professionalizing) elite interested in understanding a larger and more diverse world than had previously been imagined.[35]

The New World discoveries affected the European mind in art, philosophy, and politics in a way that directly laid the foundation for nineteenth-century environmentalism. While the ancient classical world inspired artists to depict the human body in a natural way, so too did the new discoveries feed into the Renaissance tendency to critically examine the whole world outside the human mind. This new critical examination questioned many of the key assumptions that underlay the old medieval society.

At first artists depicted the New World with a paucity of detail. Perhaps because of this, Europeans learned of the flora and fauna of the New World primarily from text and not from pictures. The accounts of Peter Martyr, Giovanni Battista Ramusio, and Richard Hakluyt, among others, are early examples. Christopher Columbus described the natural products of the New World with little detail, lacking sophisticated categories or individual names. He wrote:

I went forth with my captains and crews to see the island and if the others we have seen were as beautiful and green and gentle, this one is much more so; with large groves and deep foliage ... the singing of the little birds is such that it seems that one would never depart hence. There are flocks of parrots that obscure the sun, and other birds large and small, of so many kinds different from ours that are so wonderful. I am very certain that they are of great value, I shall take home some of them as specimens, and also some of the herbs.[36]

His descriptions were necessarily vague, as were many of the letters from colonialists back to Spain and to Britain. This vagueness is not surprising given the fact that a mass of new plants and animals presented themselves to the colonialists. Artists depicted Indians with little detail, not unlike the wild people of the forests of Europe. But while painting in Europe at the time of Columbus' discovery of America often looked inward to create a mood or a religious feeling, great painters like Caravaggio in Italy and van Dyke in The Netherlands would soon pioneer the depiction of the outward world with startling realism.

In narrative terms, the European expansion made impossible the description of the world in one volume. A popular fifteenth-century book, *De proprietatibus*

---

[35] Raymond Phineas Stearns, *Science in the British Colonies of America* (London, 1970), 51.

[36] Cited in E. P. Richardson, *Painting in America: the Story of 450 Years* (New York, 1956), 43.

rerum by Bartholomaeus Anglicus, described in one volume the whole operatic theater of the universal mechanism from the deity to the smallest creatures, including the elements, sickness, stars, history, plants, minerals, and attributes like color and texture. This handy compendium of all reality became obsolete with even the first travel narratives that emphasized new and novel particularities, disturbingly throwing many accepted generalities into confusion.

As the emphasis on the particular became fashionable in Europe, an increasingly sophisticated reading audience, though small, expected more than descriptions of nature in general categories. For example, Jacob Bobart in 1640 analyzed not the entire vegetable kingdom, but new American plants in a garden in Oxford. Another enthusiast, John Smith, investigated the soils and minerals in Virginia. Other artists began illustrating profusely the new books on nature, often with watercolors in exquisite detail.

The new Royal Society in England encouraged the emphasis on the particular. It gave encouragement to the projects of gentleman amateurs and provided a helpful network of resources through lectures, publications, and meetings. Here Englishmen began the task in earnest of collecting and categorizing specimens. European philosophers could see for themselves, as predicted by Francis Bacon, the power that the new scientific method and the process of induction gave men over nature.[37]

The New World impacted political philosophy as well as art. The reports that reached Europeans about American Indians living in a state of nature fired the imagination of political philosophers like Hobbes, Locke, and Rousseau. Without "civilization," American Indians offered insight into human nature in the wild and living examples of "natural" forms of self-governance, sometimes positive, as with Rousseau, often negative, as with Hobbes.

Hobbes saw that "the state of nature is the situation that would prevail if mankind were made up of people able to slay one another, destitute of all mutual fondness, strangers to any contractual agreements and political institutions."[38] Locke shared this rather negative view of human nature, arguing that civil society arose from the "inconveniences" of natural society and from the evils that one human inflicts on another. Citizens in a commonwealth remedy this situation by contracting one with another.[39]

Locke also laid the foundation for a philosophy of land use in the colonies, later espoused by John Stuart Mill as clerk in the East India Company. Locke stipulated that use determines the ownership of land, particularly the intensity of use. Unlike the Spanish conquistadors, the British felt the need to justify the acquisition of land in the New World and to lay a legal foundation for ownership that satisfied an objective sense of justice and satisfied the ethical right to colonize.

---

[37] For the Englishmen involved in analyzing plants and animals see Stearns, *Science in the Colonies*.

[38] David Gauthier, *Perspectives on Thomas Hobbes* (New York, 1988), 110.

[39] Scott Alexander, *Locke* (New York, 1969), 77.

Locke argued that since "God has given the Earth to the children of Men, given it to Mankind in common,"[40] then *use*, not conquest or violence, defined the right of ownership. Since society had no transhistorical legal genealogy of ownership, then proper exploitation remained the only arbiter of ownership. Generally this meant agricultural cultivation.[41] Cultivation and enclosure guaranteed ownership. In consequence English settlers who cultivated and enclosed land, not the American Indian skimming large tracts of land for sustenance, owned land. Crossing a large tract of land in search of game established nothing. "Industrious and rational title" belonged to those who labored, Locke argued.[42]

This rationale justified the seizure of land from those engaged in a semiagricultural or a hunter-gatherer lifestyle. As discussed later in this book, it also came to be extended to the public ownership of land. The Dalhousie Forest Charter of 1855 extended the idea of absolute ownership of private land to absolute ownership of public land. It also stipulated that the government upgrade the intensity of use through forestry and better utilize the land for the public good. Management of nature through the reservation of forests came to be viewed as a better handling of the public interest than nonmanagement, or *laissez-faire*. Though initially applied to forests, this argument is the key philosophical justification for the reservation of wetlands, river basins, wilderness, and other natural areas. This modified Lockean position is still the operating assumption of environmental legislation today in most parts of the world.

To Rousseau the New World cast European political society in a new light. Civilization, not the state of nature, he observed, caused human ills. The vision of native peoples living in a wilderness attracted him, and he imagined a world without property and without the conflict that property effected. Living without the artificial corruption of civilization, expressing impulses in the theater of nature, led to the amelioration of individual and societal ills.

The empire that governed this New World inspired a variety of visions for reform. Though Hobbes viewed human nature pessimistically and Locke with the optimism that accompanies a human nature malleable to education and reform, both have been characterized as holding an "imperial" view of nature, one that demands the manipulation of the natural world for the benefit of society. Rousseau, as an early contributor to Romanticism, has been characterized as holding an "arcadian" view of nature, escaping the imperial taint.

Scholars focus on two eighteenth-century men in particular, who represent an "arcadian" and "imperial" view of nature. Donald Worster argues that Gilbert White embodied the arcadian and Carl Linnaeus the imperial approach to nature. The arcadian view, Worster writes, advocated a "simple, humble life for man with the aim of restoring to him a peaceful coexistence with other organisms." It is a pagan-inspired mode, most definitely not Christian, and seen in men like White,

---

[40] As quoted in Barbara Arneil, "The Wild Indian's Venison: Locke's Theory of Property and English Colonialism in America," *Political Studies* 44:1 (March 1996): 61.

[41] ibid., 62.   [42] ibid., 63, 66.

2 Reserved teak and evergreen forest in Burma. Empire foresters saw the preservation of the forest undergrowth as necessary for the health of the soil. 1902.

culminating in Henry David Thoreau. Both White and Thoreau pressed man to "accommodate himself to the natural order rather than seek to overwhelm and transform it."[43]

The imperial tradition, according to Worster, is characterized not by a pagan but by a Christian character, decidedly imperial and scientific in nature. Carl Linnaeus fits this model because he classified and named plants, dividing nature into identifiable parts. He applied a strict Baconian and scientific method to the world. This arbitrary project of division and cataloguing only elevated humans above nature and gave society more control over the nonhuman world. Linnaeus also practiced a lively Christian faith that reinforced his belief that man had been made to govern nature.

Some problems crop up with this interpretation, however. Not the least is Gilbert White's position as a clergyman in the Church of England. He believed as passionately as any eighteenth-century scientist that God gave man dominion over nature, and that humans should never "lose sight of utility" in the study of nature. Linking White with pagan culture is difficult also because White cherished his priesthood and the work of the parsonage. Never did White doubt that nature existed in a Newtonian universe under the laws of a very Christian God.

[43] Worster, *Nature's Economy*, 2, 76.

Worster's construction of a pagan White is based more on suggestion than evidence. Worster points out that his village sponsored country dances and that White read the classics. It hardly needs a refutation that most clergymen from Oxford, by no means pagan, read the classics, and that a country dance is not unknown in Christian Europe. Such picturesque imagery as a church and a tree growing close together "rooted in the same earth" symbolize to Worster a pagan unity of man with nature. That such imagery can just as easily symbolize a connection between Christianity and nature is never examined.[44]

In fact, Gilbert White announced that the "GOD OF NATURE" (White's capitalization) expressed his wisdom through the Creation. This doctrine fits very well with the natural theology of eighteenth-century clerics. White also, in line with the growing fascination of the picturesque, lamented the absence of church spires in Selborne, for "such objects are very necessary ingredients in an elegant landscape."[45] Perhaps his position as a clergyman heightened his appreciation of nature because the church gave him the income and the leisure to pursue studies of his environment.

More difficult to correlate with Worster's portrayal of White as a pagan arcadian is the relationship that White had with animals. White determined to exterminate insects; he hated snakes; he enjoyed hunting; he tamed animals for the benefit of humans. He expressed joy when interesting animals were shot and brought to his study for observation. "I was much pleased to see a male otter brought to me weighing twenty one pounds, that had been shot on the bank of our stream," he wrote in one instance of many. He expressed relief when an "annoyance ceased" by the shooting of a pair of owls that took residence in his dove house. White did love nature, and he certainly had a strong dose of Romanticism, but not to the exclusion of Christian and utilitarian views.[46]

White did make a powerful contribution to the study of nature. He advocated ideas that we now identify as key to the environmental movement, such as the link between all aspects of nature and the mutual dependence of each part of animal life to the whole. His observations on the beneficial role played by earthworms are well known to generations of admirers. The very slogan that Worster identifies as the "imperial" position of the agricultural reformer Arthur Young, "Make two blades of grass grow when one grew before," is taken from White, who in turn took the phrase from *Gulliver's Travels*. Richard Mabey comments that English settlers packed White's *Natural History of Selborne* alongside family Bibles when emigrating to the colonies. Settlers certainly did not see Gilbert White as a hindrance to imperialism.[47]

The imperial model also has problems. This model conflates Christianity, imperialism and science into one nature-slaying ogre. Francis Bacon, perhaps the first

---

[44] ibid., 2, 4.   [45] Gilbert White, *The Natural History of Selborne* (Boston, 1975), 55.

[46] ibid., 47, 55, 71–72.

[47] Richard Mabey, *Gilbert White: a Biography of the Author of the Natural History of Selborne* (London, 1986), 6.

3 Teak tree in northeast India. 1898.

representative of the scientist, was nominally Protestant and certainly less religious that Gilbert White. But Christian, as Worster asserts? In *The New Atlantis* Bacon presents a utopia run by scientists where the Christian priests kneel before the scientists, where "The priest . . . reveres the scientist in place of bishop, pope, and perhaps even the Lord." He is credited by many as an early apostle of secularism and as the voice of sanity against superstition.[48]

Though a scientist, Bacon also perceived the ultimate reality of nature as a single entity. He suggested that the god Pan, "as the very word describes, represents the universal frame of things, or nature," and that "the office of Pan can in no way be more lively set forth and explained than by calling him the god of hunters." From this poetic vision he moves to the practical and suggests that "every natural action, every motion and process of nature, is nothing else than a hunt. For the sciences and arts hunt after their works, human counsels hunt after their ends."[49] Bacon

[48] Robert K. Faulkner, *Francis Bacon and the Project of Progress* (Lanham, 1993), 252.
[49] Bacon, *Works*, vol. VI, 709, 711.

4 Reserved evergreen forest of deodar, spruce, and blue pine in the Punjab, India, looking toward the Himalayas. Photo taken by R. S. Troup, 1914.

5 Oxen pulling a log in Burma. Oxen and elephants are the main method of transportation for logging operations in Burma and India even today.

uses language that is romantic, pagan, and arcadian, as well as at times Christian, secular, and "imperial."

Linnaeus is an odd figure to represent an imperial tradition. While Linnaeus believed in the Christian faith, he did not share the "imperial" stance of domination over nature, nor did he pursue knowledge purely for utilitarian purposes. Named by his clergyman father after a Linden tree, Linnaeus lovingly described his travels throughout the natural world, such as his trip to Lapland where his prose rivaled Thoreau's in poetic beauty. He valued every component of nature, tending to see nature as a whole where "nothing is made by Providence in vain, and . . . whatever is made, is made with supreme wisdom."[50] Linnaeus believed in a "just proportion" among species. Providence aimed to "prevent any one of them from increasing too much, to the detriment of men, and other animals." He emphasized the "reciprocal uses" of the economy of nature, overseen by God where "natural things . . . are fitted to produce general ends, and reciprocal uses." An arcadian notion not far removed – if at all – from symbiotic interactions and ecological relations.[51]

Both Gilbert White and Carl Linnaeus are better characterized as Christian naturalists. Placing them at opposite ends of a spectrum is a disservice to both men. White admired Linnaeus' classification system, which he handled "with propriety," and thought Linnaeus' "characters and genera . . . clean, just and expressive."[52] Both saw nature as an expression of the character of the Christian God. As Knut Hagberg, biographer of Linnaeus, comments on the discussion of White and Lineneaus, White emulated Linnaeus and prided himself on giving "precise details of Linnaeus' important system." Without Linnaeus, "Selborne is superfluous; it [the Linnaean system] is evident on every page." No clear dichotomies can be constructed; neither man is clearly an arcadian pagan or a cold imperialist in their approach to nature.[53]

But if environmentalism and imperialism have a shared past, a question still remains about empire forestry. Clearly empire foresters participated in and, as will be shown, supported imperialism. But did empire forestry and the conservation project it spawned in the nineteenth and early twentieth century express ideas that can be identified as "environmental" or "ecological"? To answer that question it is necessary to examine the rationale that foresters and legislators used for the intervention that resulted in the massive reservation of land.

Climate theory: the rationale for intervention

The optimism that humankind could and should change the environment for the better merged in early environmental thought with a deep premonition that things could go horribly wrong. Thomas Malthus contributed to this "managed

---

[50] Linnaeus, *Travels*, ed. David Black (New York, 1979), 16; Worster, *Nature's Economy*, 100.
[51] ibid., 99, 31.   [52] White, *Selborne*, 88, 76.
[53] Knut Hagberg, *Carl Linnaeus* (London, 1952), 160–161.

pessimism" by demonstrating in his classic *Essay on Population* that civilization produced problems as well as solving them. His inquiries – critical of the Marquis de Condorcet and William Godwin – etched the limits and possibilities of progress by eschewing notions of the perfectibility of society inherent not only in Rousseau but also in Adam Smith and the ideal of the unregulated marketplace. Malthus' formula that "population increases in a geometrical, food in an arithmetical ratio" served as a corrective to blind optimism.

Change does not always equal catastrophe, but ideas of change certainly fed into a Victorian apprehension that man could interfere with nature for the worse. The idea of nature changing, not slowly but catastrophically, caught the imagination of George Perkins Marsh, whose book *Man and Nature* alerted a worldwide audience to the dangers of deforestation. However, four key thinkers and writers directly influenced the Victorian mindset about change and nature. James Hutton, J. B. Lamarck, Charles Lyell, and Charles Darwin all discussed a changing nature, in ways that proved unsettling and substantiated the warnings of Marsh.

James Hutton argued in the late eighteenth century that the earth had neither a beginning nor an end, and that the job of a geologist consisted of tracking the slow and steady changes wrought over immense periods of time. His theory provided the groundwork for a theory of organic evolution and reinforced a cyclical view of history.[54] The French philosopher J. B. Lamarck also contributed to the idea of a changing nature when he proffered evolution by transmutation. Both challenged orthodoxy: Hutton posed a challenge to the biblical scheme of time; Lamarck to the fixity of species from a creator God.

The most broadly accepted theory in the early nineteenth century envisioned a series of discrete creations. Charles Lyell, a widely read geologist, opposed the Lamarckian view that life had an "indefinite capacity of varying from the original type." Lyell argued that new species appeared, often died out, but did not change. However, he agreed with Hutton and Lamarck that geographical features changed via observable forces such as wind, water, and lava flows.[55]

One question did not occur to Lyell. His acceptance of geological evolution, his view that nature involved a struggle where man inadvertently "reduce[d] the natural order to a smaller number of species," involved, like the geological history of the earth, unrelenting change. This affected not only natural features, but humankind.[56] But for the better or worse? Others, building on his work, asked this question.

Darwin's ideas owed much to predecessors such as Malthus, Lamarck, Hutton, Lyell, and among others, his grandfather. From Lyell he learned about the inevitability of extinctions and the formula that "large-scale features are to be understood in terms of the summation of small-scale events." Inevitably this affects, over time, species and their adaptation to their ever-changing environments.[57] Victorians

[54] James Hutton, *James Hutton's System of the Earth, 1785; Theory of the Earth, 1788; Observations on Granite, 1794* (New York, 1970), ix, xi.
[55] Peter J. Bowler, *The Environmental Sciences* (London, 1992), 19.
[56] Charles Lyell, *Principles of Geology* (Chicago, 1990), 11–31.   [57] ibid., xxx.

progressively applied the idea of change of geological formations, to plant life, to animals, and then to humans. Darwin effectively accomplished the latter.

The implications were frightening, even to Darwin. All things change, with little but the laws of nature fixed. This change involved, in fact demanded, extinction. Malthus proved a pivotal influence as well, injecting the pessimism of competition by arguing that humans would breed faster than the food supply. While Lamarck emphasized adaptation as a key to change, Darwin emphasized the power of selection as the key to adaptation. Darwin then applied this reasoning to the whole animal kingdom and concluded that the result would be a powerful force of selection. Thus "natural selection can act only by the preservation and accumulation of infinitesimally small inherited modifications."[58] When passed on to survivors, new species evolve. Darwin, unlike Lamarck, explained the origins of animal adaptations more to the satisfaction of his contemporaries. He argued that in nature "the slightest difference of structure or constitution may well turn the nicely balanced scale." Darwin helped many Victorians to understand that change interjected by humans interrupted established niches established by natural selection.[59]

Fear of human-engendered catastrophe focused early on the forests. Revenue from timber, along with the very practical concern for soil erosion, flooding, and rainfall all fed into the fear of catastrophe. To understand the implications of deforestation, to protect human society, along with flora and fauna, required highly trained experts. Scholars have defined environmentalism as a cluster of ideals arising from modernization that saw humans dependent upon proper management of the environment by scientifically trained men.[60]

Just as disease, slums, crime, and factory conditions forced Victorians into government intervention, so too did timber shortages, soil erosion, and the fear of devastating climate change. Imperial officials, along with Marx and Darwin, credited Malthus for sparking their insight. While Enlightenment optimism encouraged the remaking of the world, the attempt to define the limits of optimism – described by Malthus as resource scarcity – served to encourage a vast intervention for preservation. Malthus demonstrated that catastrophe followed hard on the heels of resource depletion. Optimism and pessimism proved a mixed drink of powerful potency. Victorian imperial officials were convinced that civilization produced problems and, simultaneously, the means to solve them. These same officials experienced the problem, prescribed the solution, and persuaded their masters, both governments and people, to take the medicine.[61]

---

[58] Thomas F. Glick and David Kohn, *Darwin on Evolution: the Development of the Theory of Natural Selection* (Indianapolis, 1996), xi.

[59] ibid., 176–177.

[60] It is particularly difficult to trace the influence of an idea. The actual impact of Romantic and pantheistic literature on the production of late nineteenth- and earlg twentieth-century scientific work – particularly ecology – has not been adequately explored, though Donald Worster makes a good beginning in his book, *Nature's Economy*.

[61] For an interesting presentation of urban reform movements, and how this impulse for reform affected the broader environmental movement, see Gottlieb, *Forcing the Spring*. Gottlieb argues that the

By the late nineteenth century dire social prophecy, mixed with optimism, abounded at all levels: economic, racial, religious, imperial, and environmental.[62] In literature, Matthew Arnold exemplified the foreboding felt at the decline of religion.[63] In social thought, Henry Mayhew and Charles Booth, along with William Morris, chronicled the growing hardship of the poor and feared (or predicted, as did Marx) the social consequences. Joseph Chamberlain expressed widespread popular concerns about Britain's fate in a world of giant empires. And the fear that natural resources would run out affected government officials in various departments.[64]

The fear of catastrophe haunted empire foresters. The Muslim conquest of the Middle East and of India served as a prime example of catastrophe to foresters who observed denuded hills in Muslim areas. Shifting cultivation – the burning of forests to clear land for a single crop of grass – and grazing for vast herds of cattle, sheep, and goats precipitated the worst damage. Berthold Ribbentrop, Inspector General of Forests in India from 1884 to 1899, wrote:

that the wholesale destruction of forests had the most deteriorating effect on the climate of India is certain, and sufficient proof of this assertion can be found in the numerous deserted village sites and mounds, indicating the previous existence of a dense population in parts of the country where cultivation is at present found only in the most favourable situations.[65]

The existence of "settlements on a scale unmistakably indicated by extensive ruins" and the evidence of ancient systems of artificial irrigation proved the case to British administrators that "the wholesale and continuous firing of the forest vegetation of the country" destroyed previous civilizations.[66]

preservation of nature outside of the cities in the United States symbolized an escape from the disease and pollution of urban areas. He gives a particularly fascinating account of Alice Hamilton, who worked as an urban environmentalist in the 1920s. For more on Hamilton see Barbara Sicherman, *Alice Hamilton: a Life in Letters* (Cambridge, MA, 1984) and "Working it Out: Gender, Profession, and Reform in the Career of Alice Hamilton," in Noralee Frankel and Nancy S. Dye, eds., *Gender, Class, Race, and Reform in the Progressive Era* (Lexington, KT, 1991), 127–147.

[62] See Dietmar Kamper and Christoph Wulf, eds., *Looking Back on the End of the World* (New York, 1989); Greg Myers, "Nineteenth-Century Popularizations of Thermodynamics and the Rhetoric of Social Prophecy," in Patrick Brantlinger, ed., *Energy and Entropy: Science and Culture in Victorian Britain* (Bloomington, 1989).

[63] See Matthew Arnold, *Culture and Anarchy* (New York, 1941), and Matthew Arnold, "Dover Beach," in *The Poems of Matthew Arnold* (Oxford, 1913).

[64] See, for one example, William Stanley Jevons, *The Coal Question: an Inquiry Concerning the Progress of the Nation and the Probable Exhaustion of our Coal Mines* (London, 1866). By the turn of the century this catastrophic prophecy operated within a larger assumption that the world was a complete and unified structure, and one which could come to an end. Martin Heidegger placed modernity within this first operating assumption. H. G. Wells and Oswald Spenglar were also early twentieth-century writers who assumed a world picture that emerged from a world capitalist system and from imperialism. But if the world could be one, it could also end as one from environmental catastrophe.

[65] Berthold Ribbentrop, *Forestry in British India* (Calcutta, 1900), 37.   [66] ibid., 37, 38.

This combination of historical imagination and eschatology produced the ruling theory of environmentalism for the nineteenth century. Government officials envisioned a massive environmental regeneration to restore the ancient and pristine balance and, to use a modern phrase, *terraform* the terrain. Ribbentrop argued that "If it once be accepted that the climate of India would regain its pristine state by a complete afforestation of a large proportion of the country, every step in that direction must exercise small advance (however immeasurable) in that direction."[67]

Contemporary reports of Indian forest administrators, scientific journals, and popular magazines kept climate theory current with the reading public and with legislators. Nineteenth-century sources prior to 1855 show a lively debate on the influences of deforestation on climate, with Europe as well as the United States holding a major place in the observations. Often reports ricocheted through a number of sources, with no one paradigm – such as Grove's "island thesis" – at work.

As an example, a report first published in Warsaw in 1826 detailed the role that the forest of Bialowieza played in water flow and climate.[68] The forest, situated in Lithuania and the favorite hunting ground of the czars, contributed more to the river flow of the Narew and Bug rivers than did high mountains or glaciers, the article claimed. The *Magazine of Natural History* reprinted the report in September 1835.[69] It gained notice again in 1845 when the book *Geology of Russia in Europe and the Ural Mountains* reported its findings.[70] Roderick Murchison, author of the work, argued after reviewing the report that

The hands of man ... are still effecting considerable changes, in large tracts of Russia, by the destruction of her forest ... a few centuries only have elapsed since northern Russia was a dense virgin forest ... but now her gigantic pine trees are felled, lakes and marshes are drained ... and in great measure account for the diminution of ... the waters of the Volga and ... the cause of increasing drought.[71]

The Proceedings of the Zoological Society of London in 1848 further discussed the report.[72] It then came to the attention of the Royal College of Surgeons and also to a Colonel Jackson, secretary of the Royal Geographical Society of London, who read portions of the report in an address to the society.[73]

[67] ibid.

[68] Baron de Brincken, *Mémoire descriptif sur la fôret impériale de Bialowieza en Lithuanie* (Chez Glucksberg, 1826), 127.

[69] *Magazine of Natural History, and Journal of Zoology, Botany, Mineralogy, Geology, and Meteorology* (London, 1835).

[70] Sir Roderick Impey Murchison, *The Geology of Russia in Europe and the Ural Mountains* (London, 1845).

[71] ibid., 578.

[72] Zoological Society of London, *The Proceedings of the Zoological Society of London* 3 (London, 1848), 12, 13.

[73] W. B. Clarke, "Effects of Forest Vegetation on Climate," *RSNSW* 10 (1876).

The phenomenon of tropical islands and deforestation was still hotly discussed in the nineteenth century. In 1830 the Reverend Landsown Guilding of St. Vincent presented a paper where he subjoined an account of rainfall measurement between 1825 and 1829. He announced – as new knowledge – his theory that climate and deforestation were linked, and he did so without any previous reference to eighteenth-century observations. He wrote that "Climate has been considerably affected by the continued industry of man and his daily encroachment on the primeval forest . . . so much has this change been felt, that laws have been passed to prevent the cutting down of timber in certain directions."[74] This tortuous path is typical of the observations of deforestation and climate change that circulated around the colonies, Europe, the United States, and elsewhere.

Another broadly circulated book, *The Plant: a Biography* concentrated its observations not on tropical islands but on the Mediterranean region.[75] The author, Mathias Schleiden, observed that clover, which required plentiful rainfall, had passed from Greece to Italy and then to southern Germany where, due to deforestation and dry conditions, it grew only in the more northern areas. He warned that

If the contended clearing and destruction of forests is at first followed by greater warmth, more southern climate, and more luxuriant thriving of the more delicate plants, yet it draws close behind this desirable condition another which restrains the habitability of a region within as narrow as, and perhaps even narrower limits, than before.[76]

He concluded with a statement reflecting a remarkably modern environmental sensibility:

A broad band of wasteland follows gradually the steps of civilization. If it expands, its center and its cradle dies, and on the outer borders only does one find green shoots. But it is not impossible, only difficult for man, without renouncing the advantage of culture itself, one day to make reparation for the injury he has inflicted; he is the appointed lord of creation.

Nature lay before man, "in her wild and sublime beauty," but behind man lay "a desert, a deformed and ruined land" because of this thoughtless destruction of "vegetable treasures."[77]

Middle-class readers also came across climate theories. Dr. Tristram's popular *Natural History of the Bible* claimed deforestation led to climate change and backed

---

[74] ibid., 181–202. Also see Charles Daubeny's lectures from the 1850s and 1860s on the effect of deforestation on climate in *Lectures on Roman Husbandry* (Oxford, 1857) and *Essay on the Trees and Shrubs of the Ancients* (Oxford and London, 1865). The rainfall of St. Helena is still discussed in 1849. See the Bombay Geographical Society's Report (1849). No island paradigm is operating or settled in the nineteenth century, but rather an observation of climate and forest relationship that is broad-based, contemporary, and international.

[75] Mathias Jacob Schleiden, *The Plant: a Biography*, trans. Arthur Henfrey (London, 1853).

[76] ibid., 304–306.   [77] Clarke, *Climate*, 41.

up his assertion by using ancient literary sources to compare the rainfall of ancient Israel to that of modern Palestine.[78] He wrote that

There is every probability that when the country was well wooded and terraced, and those terraces clad with olive trees, the spring rains were more copious than at present. Many light clouds which now pass over from the west would then be attracted and precipitated in rain over the highlands. At present, without any effort to utilize the bountiful supplies of Providence, three-fourths of the rainfall are wholly wasted.[79]

Later in the century, such observations would inundate journals, newspapers, and even general encyclopedias. Another popular book in the 1870s entitled *Information for the People* argued that "science has proved" forests exercise a "benign influence on the climate" as well as on the health, security, and prosperity of the country. New forestry methods are now practiced not only in Britain but "also her Colonies and the Indian Empire."[80] Journalists also picked up the point. The *Hawaiian Gazette* was not unique when it made the case in 1876 for forest preservation laws based on a climate theory – not of the tropical islands – but of "various European nations," where the forests are the "life of a land, as lungs are to the life of the animal. When a land is shorn of its forests, its green fields became barren wastes, its rivers became dry in summer, and rain destructive torrents in winter. [And] it finally becomes a desert, fit only for the abode of owls and bats."[81]

North America provided popular examples for the newly popular climate theory. The *Magazine of Natural History* in 1834 observed that small rivers and streams in Kentucky had gone dry precisely in those areas where timber had been cleared thirty years earlier. In New Jersey observers noted the same phenomenon with worse results, many of the brooks having disappeared forever.[82]

The classic text for climate theory pulled together various geographical examples about deforestation and rainfall. Marsh's *Nature and Man* utilized examples from Europe, the United States, and much of the world, making only scant mention of St. Helena, and none of Mauritius.[83] Even earlier, however, Marsh had traced out the basic ideas of climate and deforestation in his observations of hillside clearings in Vermont, which, he argued, were too glaring to have escaped the notice of any "observing person."[84] He warned that treeless hillsides in Vermont could not absorb the rains, so that now water "fill[s] every ravine with a torrent and [has] convert[ed] every river into an ocean." Meadows became deserts in summer,

[78] Henry Baker Tristram, *The Natural History of the Bible* (London, 1911).   [79] ibid., 33.
[80] As quoted in Clarke, *Climate*, 51.
[81] *Hawaiian Gazette*, September 13, 1876, as quoted in Clarke, *Climate*, 62.
[82] As quoted ibid., 103.
[83] It is likely that Marsh had read Alexander von Humboldt and Aimé Bonpland who coauthored *Voyage aux régions equinoxiales du nouveau continent 1799–1804* (Weinheim, 1815), as well as *Nova genera et species plantarum*, 2 vols. (Paris, 1814, 1834).
[84] George Perkins Marsh, address delivered before the Agricultural Society of Rutland County, Sept. 30, 1847 (Rutland, VT, 1848), printed p. 18 of publication of the same name.

and seas in the autumn and spring.[85] His observations of deforestation centered on Europe, especially France after the French Revolution.

It is in this milieu of broad-based discussion in scientific journals, magazines, and popular books that the climate theory was disseminated throughout the nineteenth century. No one narrow path was responsible, neither the island deforestation cited by Grove nor the group of medical surgeons working for the East India Company. This broad-based discussion gave background to the very specific concerns of Indian administrators and led to the innovations of Indian foresters and thus to the paradigm for a worldwide environmentalism. Climate theory made it clear that it may be necessary for "the planting of a wilderness" if the needs of climate demanded it, and this perception percolated through a variety of cultural venues.[86]

Rudyard Kipling provides an example of this. Among his least-known works today, *In the Rukh*, is his first Mowgli story, now eclipsed by the success of the reworked *Jungle Book* stories. First published in *Many Inventions*, the story was republished in *McClure's Magazine* in June 1896.[87] Rukh means forest, and the idea for the story came to Kipling after a meeting with Ribbentrop, who inspired the author with tales of heroes who had the "reboisement [reforesting] of all India in [their] hands." Kipling wrote this first Mowgli story to present an ideal forester as a new Adam, ready to join governmental service in the biggest endeavor in human history – nothing less than the reforestation of the world, starting with India. I quote the story at some length because it shows not only the many meanings of the word *nature* to a nineteenth-century immigrant in India, but also how empire forestry captured the imagination of the public through such a best-selling artist as Kipling. It is interesting to note that Kipling wrote the story in Vermont and published it first in an American magazine, illustrating the circulation of empire forestry accomplishments outside the British Empire:

Of the wheels of public service that turn under the Indian Government, there is none more important than the Department of Woods and Forests. The reboisement of all India is in its hands . . . Its servants wrestle with wandering sand-torrents and shifting dunes . . . They are responsible for all the timber in the State Forests of the Himalayas, as well as the denuded hillsides that the monsoons wash into dry gullies and aching ravines . . . They experiment

[85] By no means, however, ought the government step in to protect forest areas, for public ownership is an unnecessary hardship. Mash's *laissez-faire* views would not allow for any form of governmental protection. See ibid., 3–20.

[86] Clarke, *Climate*, 37. How scientific professionals persuaded Parliament and the civil service to subsidize their interests varies with each discipline. For a broad look at the professionalization of science see M. Berman, *Social Change and Scientific Organization, 1799–1844* (London, 1978) and S. F. Cannon, *Science in Culture: the Early Victorian Period* (London, 1978). Also useful for the imperial context of science is R. MacLeod, "Of Medals and Men: a Reward System in Victorian Science, 1826–1914," *Notes and Records of the Royal Society* 26 (1971): 81–105; R. MacLeod, "The Alkali Acts Administration, 1863–1884: the Emergence of a Civil Scientist," *Victorian Studies* 9 (1965): 85–112. Harold Perkin, *The Rise of Professional Society* (London, 1989).

[87] W. W. Robson, ed., *The Jungle Book* by Rudyard Kipling (Oxford, 1987), 371.

with battalions of foreign trees, and coax the blue gum to take root and perhaps, dry up the Canal fever. In the plains the chief part of their duty is to see that the belt fire lines in the forest reserves are kept clean, so that when drought comes and the cattle starve, they may throw the reserve open to the villager's herds and allow the man himself to gather sticks. They poll and lop for the stacked railway fuel along the lines that burn no coal; they calculate the profit of their plantations to five points of decimals; they're the doctors and midwives of the huge teak forests of Upper Burma, the rubber of the Eastern Jungles and the gall-nuts of the South: and they are always hampered by lack of funds. But since a Forest Officer's business takes him far from the beaten roads and the regular stations, he learns to grow wise in more than wood-lore alone; to know the people and the polity of the jungle; meeting tiger, bear, leopard, wild-dog, and all the deer, not once or twice after days of beating, but again and again in the execution of his duty. He spends much time in saddle or under canvas – the friend of newly planted trees, the associate of uncouth rangers and hairy trackers – till the woods, that show his care, in turn set their mark upon him, and he ceases to sing the naughty French songs he learned at Nancy, and grows silent with the silent things of the underbrush.[88]

Kipling's romanticizing of "that Greek God" Mowgli, "Faunus himself," reflected the complicated late nineteenth-century view of nature. Sometimes shown as secular, sometimes as teleological or even pagan, nature is dappled with multiple meanings. Kipling made Muller, the fictionalized Ribbentrop, "The gigantic German who was the head of the woods and forests of all India, head ranger from Burma to Bombay," explain his various feelings about the Rukh. Muller drops into his camp chair "with a sigh of satisfaction as he lighted a cheroot" and surveys the jungle around him in the light of the fire, to explain in his thick German accent, that "When I am making reports I am a Free Thinker und Atheist, but here in the Rukh I am more than Christian. I am Bagan [pagan] also." But despite his "bagan" sentiments, nature remained unknowable to the Inspector General and so Muller in desperation admits, "I know dot, Bagan or Christian, I shall nefer know der inwardness of der Rukh."[89]

    If the essence of nature is unknowable for civilized man, it need not remain so for the ideal forest service employee, Mowgli. Mowgli is presented as the perfect recruit, nursed on the milk of nature herself. "An angel strayed among the wood," who spoke with a voice "clear and bell-like, utterly different from the usual whine of the native." He could disappear from sight like a ghost without a sound and then appear like morning mist, cognizant of every trick of the jungle, every inclination of bird, snake and buffalo. Gibson, a forest ranger, thought, when he first met Mowgli, that he "must get him into the government service somehow . . . he is a miracle – a *lusus naturae*." Muller observed that "he is before der Iron Age, and der Stone Age. Look here, he is at der beginnings of der history of man – Adam in der garden . . . he is older than . . . der gods."[90]

    The naked Mowgli, standing godlike before a blazing fire, "the very form and likeness of that Greek God" Apollo, is told by Muller that he commanded 5,000

[88] ibid., 327.   [89] ibid., 344.   [90] ibid.

to 10,000 forest guards, and that Mowgli's new job as a forest guard is "to drive the villager's goats away" when they have no permit, to watch the game, and "to give sure warning of all fires in the Rukh." After this he would be paid and given "at the end . . . a pension." Mowgli readily accepts, for he loves the forest above all things and the rule of British forest law protected his home and playground. In youth, a jungle boy; in maturity, an empire forester.[91]

India's special role

Why did British India play a founding role in the development of environmental legislation and practice? One reason is the availability of wilderness and forest areas. Eugene Cittadino has pointed out that certain forms of scientific study occurred outside of Europe and Britain, specifically in the colonial context of European imperialism.[92] In both Germany and Britain, universities produced an oversupply of trained specialists in the field of botany, who, in the 1860s, could not all find teaching positions. Moreover, with such heavy cultivation of land in northwestern Europe, botanists could find relatively little "natural" habitats that compared with the rich diversity of the colonies. The British colonies beckoned to young graduates, who saw not a wild waste of jungle and savanna, but a frontier of new knowledge, adventure, and discovery.[93]

---

[91] ibid., 343.

[92] Eugene Cittadino, *Nature as the Laboratory: Darwinian Plant Ecology in the German Empire, 1880–1900* (Cambridge, 1990), 3.

[93] The question of the relationship between empire and science, in this case between empire and environmental science, has evolved in recent years to be not, "How did science make empire?" but "How did empire make science?" See F. V. Emery, Geography and Imperialism: the Role of Sir Bartle Frere, 1815–1884," *Geographical Journal* 160 (1984): 342–350; D. R. Headrick, *Tools of Empire* (New York, 1981); D. Mackay, *In the Wake of Cook: Exploration, Science, and Imperialism* (London, 1984); R. MacLeod, "Scientific Advice for British India: Imperial Perceptions and Administrative Goals, 1898–1923," *Modern Asian Studies* 9 (1975): 343–384; R. MacLeod, "On Visiting the Moving Metropolis: Reflections on the Architecture of Imperial Science," *Historical Records of Australian Science* 53 (1982): 1–15; M. Osborne, "The *Société zoologique d'acclimatation* and the New French Empire: the Science and Political Economy of Economic Zoology During the Second Empire," unpublished Ph.D. thesis, University of Wisconsin, 1987; J. M. Prest, *The Garden of Eden* (London, 1981); L. Pyenson, "Cultural Imperialism and the Exact Sciences: German Expansion Overseas, 1900–1930," *History of Science* 20 (1982): 1–43; L. Pyenson, "Astronomy and Imperialism: J. A. C. Ondermas, the Topography of the East Indies, and the Rise of the Utrecht Laboratory, 1850–1900," *History of Science* 26 (1984): 39–81; L. Pyenson, "*In partibus infidelium*: Imperialist Rivalries and Exact Sciences in Early Twentieth-Century Argentina," *Quina* 26 (1984): 39–81; J. Secord, "King of Siluria: Roderick Murchison and the Imperial Theme in Nineteenth-Century British Geology," *Victorian Studies* 25 (1982) 413–442; R. A. Stafford, "The Role of Sir Roderick Murchison in Promoting the Geographical and Geological Exploration of the British Empire and its Sphere of Influence, 1855–1871," unpublished D.Phil. thesis, University of Oxford, 1986; R. A. Stafford, "Geological Surveys, Mineral Discoveries, and British Expansion, 1835–1871," *Journal of Imperial and Commonwealth History* 12 (1984): 5–32. For the above suggestions I am indebted to Richard Harry Drayton, "Imperial Science and a Scientific Empire: Kew

India in particular lured specialists. The jewel in the British crown proved multifaceted, with vast forests, savanna, and grasslands studded with exotic animals and fauna.[94] Further, British investment in India's economy and infrastructure increased steadily throughout the nineteenth century, and (fortunately for "the fathers of Indian forestry") British administrators hired fairly indiscriminately among Europeans, Germans in particular, with a background in botany or forestry.

This shift toward environmental practice in the nineteenth and early twentieth centuries depended upon a shift toward state intervention and a decline, in the mid to late nineteenth century, of the entrepreneurial ideal that allowed private entrepreneurs to exploit resources without looking to the future.[95] A. V. Dicey named the 1860s as the turning point toward collectivism.[96] The Benthamite appeal to reform the new industrial cities and impose sanitation, police, education, and work standards marked a new form of state intervention hitherto unknown in liberal Victorian society. It amounted, Harold Perkin argued,

to a major change in the attitude of the State to the free market, from the assumption that the market could be safely left to the hidden hand of self-interest and competition . . . to the assumption that, although the market should still be free, the strong could not be expected not to exploit the weak unless the State laid down some very firm rules of conduct for all bargainers.[97]

Indian civil servants – who dutifully read the London *Times* – undoubtedly noticed the huge Victorian outcry against the attempt to enclose commons by private landlords. Attempts to slow the conversion of public land into private land had already progressed by the time Dalhousie issued the Forest Charter in 1855. The General Enclosure Act of 1836 banned enclosure within ten miles of London, and protected other large towns from enclosure at varying distances from the city limits. But the attempt by landlords to enclose Hamsptead Heath, Clapham, Plumstead, Tooting, Graveney, and Epping Forest aroused fierce public opposition in the 1860s. By 1865 the Commons Preservation Society was established and in 1866 the Metropolitan Commons Act provided for a public management of commons within London itself. From this point on enclosure of open fields virtually ceased.[98]

Gardens and the Uses of Nature, 1772–1903," unpublished Ph.D. thesis, Yale University, 1993, 15–16.

[94] For a glimpse of late nineteenth-century ideas concerning wildlife and conservation, see E. Evans, Ethical Relations Between Man and Beast, *Popular Science Monthly* (September 1894): 634–646.

[95] Men of letters were not far removed from this shift. For instance John Ruskin not only advocated the "natural" in art, but also advocated national education, organization of labor, and the policy of interference by the state. His essays on economics appeared in *Cornhill Magazine* (London, 1860) and *Fraser's Magazine* (London, 1862–1863) and were later published as *Unto this Last and Other Essays on Art and Political Economy* (London, 1907).

[96] See Harold Perkin, *Origins of Modern English Society* (London, 1969), 437.   [97] ibid., 439.

[98] Harold Perkin, *The Structured Crowd: Essays in English Social History* (Sussex, 1981), 110, 192–193.

Against this glacial shift from Adam Smith individualism to Benthamite collectivism emerged the new environmental interventions – paternalistic, radical, and previously untried – first in British India, then the other British colonies and the United States. This shift was praised by Bernard Fernow, chief of forestry in the United States in the 1880s and 1890s, as a necessary change "since the propaganda of forestry began." He argued that "Socialist attitude" led to an expansion of federal power in various directions, of which environmentalism claimed its share.[99]

The age of imperialism could easily be dubbed the "age of the great interference." Nature, increasingly in the nineteenth century, could be interfered with and improved. This radical concept required both eighteenth-century ideas of progress and nineteenth-century scenarios of imminent doom. The only question remaining for true believers in progress was this: is nature to be left alone as an autonomous mode of production, or is it to be managed, and possibly improved? Both answers were proposed, but the latter predominated. Mixed with the elixir of imperialism, the greatest interference the world had ever witnessed began to unfold. Vast tracts of land were claimed for the state, declared protected areas, and invaded by fireguards, rangers, and administrators. Utterly unlike the conservative royal forestry of French and English kings, the new environmentalism took action to expand and enforce a multiuse model of management over areas new to European rule. Armed with a western conception of law and a formula of absolute property rights for both the individual and the state, forestry became an international profession with global specialists ruling an empire of trees and grasslands. The newly protected forests marked the formalization of a divorce that had been threatened (as Glacken argued) since the early modern period – the final separation of man *in* nature, to man *over* nature.

---

[99] Bernard Fernow, *A Brief History of Forestry in Europe, the United States, and Other Countries* (Toronto, 1913), 159–160.

3

Empire forestry and British India

Forest administrators in India linked forest cover to climate. By the 1850s "it can hardly be denied that the existence or non-existence of large well-wooded areas in a country naturally capable of growing forests affects its climate in a very marked degree."[1] Thus precipitation depended upon forests as well as the evaporation of water from the world's oceans. A forest, whether evergreen or deciduous, made the difference between lush green countryside and sandy waste. Administrators in India saw both teak plantations and the reservation of existing forests as essential to the preservation of good climate.

It is possible to misunderstand Indian teak plantations as the farming of timber with little diversity or ecological value. This is a mistake. Early Indian foresters gave careful instructions to guarantee that teak forests were not "pure forest of large extent." Since the teak tree is leafless during the dry season, the soil loses shade during the hottest part of the year. Thus the soil is easily washed away in the monsoons. Accordingly, a mix of trees is advisable for the teak plantation, with trees that provide shade and also allow for a variety of undergrowth. Vines and underbrush protected plantation soil in this capacity as well. Working plans of plantations involved thinning out teak trees, planting a variety of species, and monitoring soil and water quality.[2]

Plantations reclaimed desert or waste areas where man's abuse had destroyed a once pristine "household of nature." The reclamation focused on the reestablishment of forests, ground cover, grasses, and other foliage that foresters thought equally important in the absorption of moisture. The term used by foresters themselves was *arboriculture* – the cultivation of trees. Arboriculture was not confined exclusively to the forest department in the nineteenth century. The canal departments in numerous provinces also planted large plantations of forests along canals, reservoirs, waterways, and roads, including areas found in the hills where water was held and dispersed by forest areas.[3]

---

[1] Ribbentrop, *Forestry*, 40.   [2] ibid., 201, 204.   [3] ibid., 200.

Dietrich Brandis, India's first forest chief, strategically pressed the practice of shifting cultivation into the service of ecological terraforming. He rushed the implantation of seedlings grown in nurseries into the soil after a burn. This turned a particularly damaging practice into the cultivation of forest cover, and necessarily kept plantations small but still provided excellent income to the forest department. Plantation extension grew annually, slowly adding thousands of acres per year to the productiveness of large forests. The annual extension of forests constituted another environmental innovation, since the annual extensions were considered "indefinite" and did not stop until the British left India in 1947.[4]

Imperial officials also extended forest cover by demarcating waste or desert lands deemed suitable for forests, a less intensive method of arboriculture, albeit one that still involved policing and governmental involvement. Foresters followed the clues of nature for this action, recommending demarcation only where a forest had once covered an area. Ruins of old cities projecting from the desert sands often gave the clue that rainfall had once been higher. Officials then marked the new area as protected in order to allow forest to reseed and grow without artificial intervention. This latter process officials found both economical and practical, since many areas of northwestern India could be reforested without intervention. It allowed the department to then focus on grazing control and fire prevention, a far more economical task.[5]

But how do we connect the vast "terraforming" of the Indian subcontinent with ecological regeneration? The answer is by listening to the foresters involved and analyzing the nature of their innovations. The *Indian Forester* is replete with examples of "forestry" concerns that reflect the "household of nature" concept. The concerns range from the quality of the soil, air, and water, to game, birds, fungi, insects, and wild animals. Even the pollution of water by factories – particularly mills and logging operations – was entailed in this concept of "forestry." The forest was considered the "Great Laboratory" where "Romance was found in the soil when it ceased to be thought of as dead mass and was discovered to be living, moving world."[6] The forest held a universe of microscopic life, and affected climate, spiritual harmony, health, and all life on earth.

How broadly did foresters imagine the regeneration of nature? One author in the *Indian Forester* warned, humorously, that his observations of the forests on the planet Mars (the dark spots) led him to conclude that the planet would be an entire desert (the light spots) if the inhabitants did not learn from the mistakes from humans and initiate the protection of forests as done in India![7] As will be shown in

---

[4] ibid., 197.

[5] "Forest Administration in Kashmir, from 1891–1895," *Indian Forester* 22 (1896): 104–109; "Forest Offenses and their Prevention," *Indian Forester* 33 (1907): 19.

[6] "The Great Laboratory," *Indian Forester* 33 (1907): 467.

[7] "The Forests of the Planet Mars," *Indian Forester* 33 (1907): 725. An example of forestry as broad-based environmentalism can be seen in the list of articles that follow, all from the *Indian Forester* from the years 1892 to 1907, when its influence on the other colonies and the United States stood

40   *Empire forestry and the origins of environmentalism*

the subsequent chapter, such technical concerns as fire protection, grazing policy, demarcation, and working plans were essential innovations for the protection of not only timber but also the entire "household of nature."

Foresters were convinced that at the time the British came to India the forest areas had been considerably reduced from the minimum necessary to maintain proper climate. India stood in a long line of environmental disasters brought about by Muslims and other invaders. Inspector General Ribbentrop differed in no way in his views on climate and deforestation from his two predecessors, Dietrich Brandis and Wilhelm Schlich. Dalmatia, he believed, had become a "stoney desert," and Persia, once a granary of the Middle East, now stood barren and desolate with exposed rock the prominent feature of the landscape. North Africa at one time fed the corn markets of Rome but now was a desert waste. Spain, Italy, Sicily, Greece, and Asia Minor also suffered from deforestation and were dryer than in ancient times.

The Muslim conquest served as a prime example of catastrophe in this respect. Foresters observed that hills and countryside in Islamic areas were denuded by shifting cultivation and grazing. Ribbentrop wrote that "the wholesale destruction of forests had the most deteriorating effect on the climate of India ... Sufficient proof of this assertion can be found in the numerous deserted village sites and mounds, indicating the previous existence of a dense population in parts of the country where cultivation is at present found only in the most favorable situations."[8]

at its height. See Tautha, "A Plea for our Feathered Friends," *Indian Forester* 18 (1892): 226; W. L. Schlater, "The Economic Importance of Birds in India," *Indian Forester* 18 (1892): 109; "Extermination of Wild Beasts in the Central Provinces," *Indian Forester* 18 (1892): 255; "Influence on the Vegetation of a Forest of the Removal of Dead Leaves from the Soil," *Indian Forester* 19 (1893): 132; Rahdar, "Influence of Places on Spirits," *Indian Forester* 20 (1894): 367; "The Restoration of Scenery," *Indian Forester* 20 (1894): 400; "Symbiosis and its Effects on the Planting of Forest Trees," *Indian Forester* 22 (1896): 85; "Forest Officers as Photographers," *Indian Forester* 22 (1896): 221; "For Little Known Trees," *Indian Forester* 22 (1896): 453; "Extraordinary Flights of Butterflies," *Indian Forester* 26 (1900): 513; "Destruction of Game in the C.P.," *Indian Forester* 27 (1901): 55; "The Food of Nestling Birds," *Indian Forester* 28 (1902): 97; "The Insect World in an Indian Forest and How to Study It," *Indian Forester* 28 (1902): 327; "The Habitat of the Red Junglefowl," *Indian Forester* 28 (1902): 365; F. Régnault, "Deboisement and Decadence," *Indian Forester* 30 (1904): 346; J. H. Maiden, "Where are the Largest Trees in the World?" *Indian Forester* 30 (1904): 610; "The Prohibition of Grass Burning and its Effect on the Game of the Country," *Indian Forester* 31 (1905): 301; "The Nilgilri Game and Fish Preservation Association," *Indian Forester* 31 (1905): 707; "The Royal Society for the Protection of Birds," *Indian Forester* 33 (1907): 728; "The Danger of the Formation of the Pure Forest in India," *Indian Forester* 33 (1907): 505; "British Empire Naturalists Association," *Indian Forester* 33 (1907): 368.

[8] Berthold Ribbentrop served as Inspector General of Forests in India for fifteen years, during which time he superintended the extension of the forest service established under Brandis. He is particularly lauded for making the Indian forests a paying proposition for the government of India. He published in 1900 *Forestry in India*, the standard record of the Indian forestry department and the base document for E. P. Stebbing's expanded *History of the Forests of India*. Both Ribbentrop's and Stebbing's histories, however, read like governmental reports cobbled together, privileging local and district histories over a unified India-wide narrative. For biographical information on Ribbentrop, see "The Retirement of Mr. Berthold Ribbentrop, CIE," *Indian Forester* 33 (1900): 614.

This combination of imagination and ecological concern produced the ruling theory of environmentalism for the age. Rather than destroying previous civilizations directly, it was thought that Aryans and Muslims had merely cut forests for farming and grazing. This in turn led to drastic climate change. Ruins in the desert were ghostly reminders of the need for forest protection.[9]

There were those who disagreed with this theory, however. Okar Peschel, an Indian forester and contemporary with Ribbentrop, questioned the importance of forests on the climate:

The amount of rain depends on the extent of oceans and seas, on the degree of heat, and on the rapidity with which the air moves over the surface of the waters. None of these conditions are changed by the extent or absence of forests. All air-currents blowing from the sea are year by year charged with the same amount of moisture, which precipitates as soon as the air is cooled below the point of saturation. If such precipitation be caused by forests, the air currents reach the regions behind these forests dryer and unable to yield a further supply of water.[10]

Ribbentrop argued against this position and compared a forest to a landlord who spent his income in his own district as opposed to an absentee landlord who took his wealth to another country. A forested area collected and released moisture from the very area where it rained, whereas a deforested area had so much runoff that the water transferred back to the ocean, with little absorption or benefit to the area. Assam provided an excellent example to Ribbentrop. Its broadly forested valley benefited from rain clouds that formed outside the monsoon season, when no inborn clouds arrived from the sea. Reevaporation explained how forests survived in this valley, since moisture from the spongy soil absorbed moisture from the monsoon, avoided wasteful runoff, reserved water in the plants, and only slowly released water back into the atmosphere, where it would rain down again.[11]

Ribbentrop relied upon such theorists as Ernst Ebermayer, a German botanist and physicist, who gave impressive empirical justification to Indian foresters for the prevailing climate theory. In *Die Physikalischen Einwirkungen des Waldes auf Luft und Boden* Ebermayer illustrated how

The forest alone, without the cover of dead leaves, diminishes the evaporation by 62 percent, as compared with that in the open. Evaporation is consequently 2.6 times less in the forests. A covering of dead leaves and vegetable mold diminished evaporation by a further 22 percent . . . Forests with an undisturbed covering of dead leaves and vegetable mold lessen the evaporation as compared with that in the open by 84 percent.[12]

[9] Ribbentrop, *Forestry*, 37, 38.
[10] Okar Peschel, as quoted in Clarke, "Effects of Forests on Climate," 200.
[11] See Ribbentrop, *Forestry*, 37, 38, 43; John Nisbet, "Soil and Situation in Relation to Forest Growth," *Indian Forester* 20 (1894): 3.
[12] As quoted in Ribbentrop, *Forestry*, 43. Ernst Ebermayer has been largely forgotten today. For biographical information on Ebermayer see *Ernst Wilhelm Ferdinand Ebermayer*, Kremery Reference files, biographical materials, folder 1900 9999, University of Wisconsin, Madison.

Ribbentrop concluded from these figures that more than half the rain stored by vegetated soil is reevaporated to form rain clouds. He offered a rule of thumb: "If therefore 30 percent of the country was under complete forest, the rainfall throughout should increase by 10 percent." This 10 percent would be primarily released in the dry season and make the difference between a desert incapable of supporting civilization and a territory capable of supporting agriculture and people.[13]

Ribbentrop buttressed his argument by the general anecdotal conviction of governmental workers, such as Major-General Fisher, a resident of Bellary and Raman Durg, who told a typical story.

I arrived in the Bellary District in June 1856 and visited the Raman Durg at once; the hills were then covered with a good strong jungle: there was always a heavy cloud during the night resting on the hills for the greater part of the day: rain fell during the southwest monsoon constantly and frequently; during the northeast monsoon it was much lighter; in the months of March, April, and May, the mango showers were usually very heavy and accompanied with much thunder and lightning. The average rainfall we calculated was then 45 inches in the year; all the springs about the hills ran abundantly throughout the year, and the Nareehulla, the main feeder of the Darojee tank, with all its tributaries, had water running in them all through the year. The climate of the Durg during the monsoons and the cold weather was quite cold enough to make fires very necessary, although its elevation is not more than 3,300 feet above sea-level. The water supply was most abundant during the whole of the hot weather, and the tank was almost always full, surpassing very largely during the southwest monsoon.

These observations refer to the years 1856 up to 1864 inclusive, when I left the Bellary District and did not visit the Durg again till January 1879. I found everything changed; the jungle had been almost entirely destroyed; the rainfall is most precarious, and certainly not so much as 24 inches in the year; the tank has not filled for the last three years, and is generally 10 or 12 feet below full tank level; the springs are almost always dry, dribbling only at the best; the climate is so changed that in the cold weather it is hardly necessary to shut the doors and windows; except for the high wind and the slight mists of the southwest monsoon, it would not be necessary to close the house at all. The main feeder of the Darojee tank dries up altogether by the end of February, and all its tributaries have no water in them.[14]

---

[13] Ribbentrop's observation of the effects of deforestation on climate were widely shared. See C. R. Markham, "On the Effects of the Destruction of Forestry in the Western Ghats of India on the Water Supply," *JRGS* 36 (1866): 180–195; George Bidie, "Effects of Forest Destruction in Coorg," *JRGS* 39 (1869): 77–90; J. Storr-Lister, "Tree Planting in the Punjab," *Cape Monthly Magazine* 14 (1877): 363–368; G. Grieg, "Threatened Destruction of Forests in the North-West Provinces of India: its Causes and Consequences," *JRGS* 1 (1879): 516–517; "The Effects of Forestry on the Circulation of Water at the Surface of Continents," *Indian Forester* 28 (1902): 1; J. H. Maiden, "Forests Considered in their Relation to Rainfall and the Conservation of Moisture," *JRSNSW* 36 (1902): 211–240. For an interesting but decidedly minority view against the ruling climate theory see "The Influence of Forestry on Water Supply," *Indian Forester* 18 (1892): 277.

[14] Ribbentrop, *Forestry*, 47.

The most conclusive evidence to foresters like Ribbentrop lay in the comparison between ancient and present conditions. Fa-Hien, a Chinese explorer in the fourth century, described the north of India as temperate and covered with forests, even in the northwestern section, which, in the nineteenth century contained only scrub and occasional dunes. To be sure, the forests seemed an obstruction to Indian agriculture. The growth of the Indian population strained existing farmland while the export of commodities like tea, spices, and aromatic hardwoods only accentuated the problem. The arrival of European farmers and large "plantains" (plantations) further exacerbated the shortage of good farmland. But the desire to rid the subcontinent of much of its forest cover did not arise with the British or the governors of the East India Company. The inhabitants of India had abused the forest for centuries before the rise of European imperialism. Shifting cultivation had been practiced in the tropical regions for thousands of years, and the forests were regularly fired to stimulate the growth of grass for grazing. The Aryan invaders entered India some time around 2000 BC, bringing with them the pastoral practices of the steps as well as settled agriculture. They burned and cleared away dense forest areas in order to obtain the growth of crops and to clear space for the grazing of cattle.

This process of invasion-as-war-on-the-forests is recorded in the ancient epic *The Mahabharata*. This epic mentions the burning of the Khunbdava forest between the Ganges and the Jumna rivers. The legend describes woods as gloomy and dark – with a defensive Indra pouring down frequent rains to quench earthborn fire. But the resolution of the heroic settlers could not be shaken, and large areas were cleared of arboreal obstruction. In the second epic of the *Ramayana*, when the Aryans settled in the Oudh, forests are depicted as a dense wilderness harboring only disease and danger, obstructing the growth of civilization.[15]

British foresters found it difficult to estimate the amount of forest that "originally" covered the subcontinent region from these sources, so they often referred to Greek writers. Western nations had the first description of the subcontinent with the invasion of India by Alexander the Great in 327 BC, while the ambassador to Persia, Megathenes, kept careful descriptions of the region. The records after this period are scanty, though some Chinese pilgrims who traveled to sacred monasteries in the Himalayan region give descriptions of forest cover as well.[16] In spite of the Aryan settlements, the descriptions from Greek writers at the time of Alexander the Great's invasion still showed the forests in the Punjab to be dense. Forests blanketed the country around Jhelum, and gave cover to Alexander's armies. Arrian, who later summarized the records of the Greek invasion, observed that the forest

---

[15] *Mahabharata*, trans. Pratap Chundra Roy (Calcutta, 1883–1894). For critical commentary see W. Hopkins, *The Great Epic of India* (New York, 1902); *Ramayana*, trans. M. N. Dutt (Calcutta, 1894).

[16] For a discussion on ancient India's influence on the western imagination, see S. Darian, *The Ganges in Myth and History* (Honolulu, 1978); J. Drew, *India and the Romantic Imagination* (Oxford, 1987).

tracts of the country were vast, "shrouding it with umbrageous trees of stoutest growth and of extraordinary height; that the climate was salubrious, as the dense shade mitigated the violence of the heat, and that copious springs supplied the land with abundance of water."[17] In addition the Pabbi and the low-lying valley between the range and the Chenab are described with forests (*Dalbergia sissoo* and *Acacia arabia*) in abundance.

The Muslims appear to have cleared the most forest prior to the British. Stebbing held the Muslim incursion to be the worst, for it drove back the agricultural population into the forests and mountains to make way for the stream of Muslims who took their homes and lands. The Muslims' introduction of vast flocks of goats, which ate plants down to the roots, exacerbated the problem, devastating seedlings and increasing the erosion of soil.[18] However, it is also true that Muslim emperors planted shade trees along the waysides of the imperial highways. The Emperor Akbar directed in the Sunnud, for instance, "that on both sides of the canal down to Hissar, trees of every description, both for shade and blossom, be planted." This gave a foretaste of the tree in Paradise, and gave rest to travelers between cities.[19]

The decay of Mogul power and the successful campaign to drive the French from India left the East India Company as territorial lords over much of the Indian subcontinent. By 1800 the Company sought not only to trade but also to rule. Stebbing concluded that the position of the forests at the time the British attained supremacy was but a fraction of the original cover.

No policy regarding forests existed at the time the British attained supremacy in India. Neither the East India Company nor the crown, which jointly ruled sections of India until 1857, had knowledge of tropical forestry, or of any forestry at all, for that matter. Since coal had come to replace wood fuel in Great Britain, the tradition of forestry in the British Isles had degenerated. The forests in India by the late eighteenth century were used merely for fuel and commerce, with an active trade in certain articles of wood for export: teak was the primary wood used in the construction of ships, with sandalwood a luxury article, and resins for turpentine, tar, and other forest products exploited for export. Accordingly, British rule in the eighteenth century witnessed a rapid increase in the rate of forest destruction.[20]

Concern over timber supplies increased after the loss of the American colonies. Then, with the advent of the wars against the French Revolution and Napoleon, the timber trade became increasingly international and strategic.[21] The Royal Navy relied in part upon supplies from the Cape colony. The supplies proved inadequate during the Napoleonic Wars, due primarily to deforestation in that territory.

---

[17] Arrian, *Anabasis and Indica*, trans. E. J. Chinnock (London, 1893), 212.

[18] See Edward Stebbing, *The Forests of India* (New Delhi, 1982), vol. I, 31.

[19] *Calcutta Review* 12 (1857): 47.   [20] Stebbing, *Forests of India*, vol. I, 38.

[21] C. A. Bayly, "The Middle East and Asia During the Age of Revolutions, 1760–1830," *Itinerario* 2 (1986): 80.

Malabar teak and other hardwoods became the focus of renewed interest and led to a survey in 1810 and a report on the teak supply in the Dang Bhils in 1811 by Captain Morier Williams. Further reports followed and marked a growing interest in India as a strategic supplier of woods for the defense of the empire. Empire forestry in concept, if not in name, had begun to emerge for strategic reasons.[22]

The governor's council at Fort William, Calcutta, early expressed an interest in the dwindling supply of timber. Though the discovery of abundant softwood trees in the foothills of southern Nepal alleviated the concern for some years, the presidencies of Calcutta, Bengal, and Bombay experienced acute timber shortages. A burgeoning population and the expansion of British rule both into Maharashtra on the west coast and north to the borders of present-day Nepal accentuated the shortage.[23] The search for more timber augmented the drive for imperial expansion and fulfilled the need to gain control over disruptive tribal groups and offer new forest resources. In the Western Ghats the acquisition of control over the Dang Bhils resulted in a fresh supply of timber, because not only were the forests rich in teak but, importantly, accessible to coastal waters for transport.[24]

Thus the dangers of shifting cultivation, grazing, and wholesale timber extraction were slow to be realized. By 1800 the whole policy of the Indian government mirrored other frontier societies at the time; extension of agriculture and ridding the government of land ownership as quickly as possible – by giving or selling to private individuals and companies.

In 1805, however, the court of directors received a dispatch from Whitehall inquiring into the feasibility of extracting teak from Madras due to the limitation of oak in England. The Napoleonic Wars, following the loss of the American colonies, worsened the problem of supply. Stebbing pointed to this inquiry as the first important interest taken by the government of Britain in the forests of India. The wooden walls of defense for the British Empire thus demanded, for the first time in its history, the protection of forests outside its own borders. And with the absence of oak plantations in England, the forests of the empire came to be eyed with envy and concern.

The Royal Navy knew that Arabs had imported teak from Bombay for the construction of their own fleet, military and commercial. So the directors of the East India Company set up a forest committee to make an investigation into the matter of requisitioning mature teak trees. The results of the inquiry were apocalyptic, perhaps one of the truest telltale signs of environmentalism in the modern sense.

---

[22] G. Albion, *Forests and Sea Power: the Timber Problem of the Royal Navy, 1652–1862* (Cambridge, MA, 1926) is the classic text in the field. See particularly pp. 346–369.

[23] For a discussion of the timber shortage at Fort William, see the National Archives of India, New Delhi, *Home Public Consultations, Foreign and Political Select Committee Proceedings*, no. 12, January 6 to December 29, 1767.

[24] The anxiety over timber supplies by the EIC are documented in BL, Home Misc. IOL, (102) f/4/39. Also see Grove, *Green Imperialism*, 380–399; R. A. Wadia, *The Bombay Dockyard and the Wadia Master Builders* (Bombay, 1955).

The inquiry contradicted the previous assumption of unlimited teak supplies. Those accessible forests that had provided supplies to date were exhausted of teak and, in order to tap the resources of distant forests, many of which remained unmapped, costly roads needed to be constructed. The committee recommended the protection of teak and the introduction of forestry methods along European lines.

What happened subsequently proved pivotal and innovative. A general proclamation prohibited all unauthorized felling of teak trees. Additionally, unclaimed lands were declared property of the crown, not open to any merchant adventurer willing to exploit "wastelands." Private interests could no longer merely pay a royalty to the East India Company and harvest what they pleased. Though the government did not enunciate the principle, the assumption that private interests could extract unlimited timber from forests was dead. Now the East India Company held teak trees for the crown as government property.

The home government appointed a special officer to safeguard the supplies of the King's Navy and gave him the duties to preserve the growth of teak and ensure its proper protection for shipbuilding. As if to underscore the new innovation of government ownership over unclaimed crown lands, the directors appointed a police officer, Captain Watson, as India's first conservator of forests on November 10, 1806. Then in April 1807 the home government expanded his powers and job description. Though the scope of his powers was vague, therein lay the novelty. For the first time a modern nation-state appointed an officer with vague police powers over "waste" lands unclaimed by private owners, and charged him with the preservation of forests.

Captain Watson attacked his new responsibilities with verve and energy. He established a government monopoly throughout the Madras and Travancore area; restricted severely the right of private timber interests to fell trees; limited native access to grazing in the forests to protect seedling teak; and simultaneously guaranteed cheap timber supplies for the King's Navy. Imperial exploitation and environmental protection converged.

But when the war with Napoleon ended, the crisis over scarce timber subsided. After 1815 trade with the United States resumed, and Sir Thomas Munro, governor of Madras, quashed the newfangled environmental regulations on the grounds that they interfered with *laissez-faire* principles. The merchant's reaction to the increased police power of the forest officer was, Munro knew, bitter opposition. Without regulation, "Private timber," he wrote, "will be increased by good prices, and trade and agriculture will be free from vexation."[25] Abruptly in 1823 the government of India abolished the first forest conservator in the British Empire, and once again gave private interests unfettered access to the forests.[26]

[25] Ribbentrop, *Forestry*, 65. For a detailed account of Thomas Munro's opposition to Watson and the government regulation of timber supplies, see Thomas Munro, "Timber Monopoly in Malabar and Canada," in A. J. Aburnoth, ed., *Major-General Sir Thomas Munro: Selections from his Official Minutes and Other Writings*, vol. I (London, 1881), 178–187.

[26] Stebbing, *Forests of India*, vol. I, 65.

A marked tension arose between the priorities of the navy – with its timber needs – and a *laissez-faire* colonial government. After a period of six years the Indian Navy Board recommended in 1831 that the forest conservatorship be resurrected. In response the court of directors asked the Madras Board of Revenue for advice. The board decided against action, primarily because it feared a competing governmental body. With that preference in mind, in 1842 the court of directors decided to implement not a forest conservancy, but new teak plantations to remedy the shortage. Accordingly, the Madras government appointed a revenue officer from Madras, the Collector of Malabar, Lieutenant Conolly, to inaugurate the Nilumbur teak plantations, constituting the first forest planting initiative in India.[27]

Commenced in 1845, the teak plantations of Nilambur acquired a worldwide celebrity. The plantation supplemented the 19,000-acre Nilambur Forest division and allowed the department to engage in teak-growing experiments while simultaneously preserving a wide array of other trees and flora. It also allowed the department to argue that income from the plantation could in the future help pay for the administration of other forest areas.[28]

The plantation would need management, and the court of directors appointed a scientific advisor on forest management, Dr. Gibson, who had previously served as a conservator with the Botanical Gardens in Calcutta. This small step set a significant precedence not only for India but eventually for the empire. First came the need to protect certain trees, then to establish plantations. Next came the appointment of not timber merchants but of scientifically trained men, usually from the field of botany, to manage and direct forest operations.[29] Empire forestry with Dr. Gibson gained its first prototype forester.

Dr. Gibson, a surgeon in the East India Company, also spoke fluently the main Indian languages, including Gujarati, Marathi, and Hindustani. As a vaccinator in the Deccan and Kandesh, he had gained intimate acquaintance with the biota of India and became alarmed at the effects of deforestation. He worked as the superintendent of the Dapori Botanical Garden, where the court of directors praised his application of Indian biota to medicinal drugs. He also wrote a widely circulated report to the Bombay government in 1840 concerning the dwindling teak supplies, and this led to his appointment as conservator of the forests in the Bombay presidency.

Gibson became an influential propagandist in the fight to reserve forests in India. In the contention between the advocates of *laissez-faire* and of environmental intervention, Gibson threw his weight behind the latter. After his appointment as consultant to the Bombay presidency, he corresponded with Sir William Hooker, the eminent English botanist, who encouraged him to promote the reservation of

---

[27] For an excellent local forest history of the presidency of Madras, see "Forestry in Madras," *JSA* 49 (1901): 757–784.

[28] Brandis and Smythies, *Forest Conference, 1875*, 95.

[29] "Botany and the Forest Department," *Indian Forester* 26 (1900): 56, 245.

forests by the state.[30] Gibson warned Hooker in March 1841 that "The Deccan is more bare than Gujarat" while the ghat mountains contain trees only in the clefts "and even there they are disappearing fast."[31] He asked for Hooker's influence to persuade the government to assume control of the forests in the Deccan and Konkan region. He also urged that since the amirs of this region had charged natives for forest products they had extracted, the new imperial landowners could and ought to consider the collection of revenue from forests as well.[32]

The argument proved alluring. The Bombay government petitioned the government of India to set up a conservancy in Bombay with Gibson as conservator of forests, all to be financed – with profits – by the revenues made from the sale of trees thinned from the forests. The Indian government agreed and appointed Gibson conservator of forests. State forest management, the first of its kind in history, had been established.[33]

The Madras and Bombay conservators inspired emulation in Burma. But the situation in Burma had a different legal history. The exploitation of teak forests by royal fîat occurred long before the British occupancy, and this had proved advantageous to the Burmese royal government. In Tenasserim the teak had been proclaimed a "royal" tree by the Alompra dynasty. When the Tenasserim Provinces were ceded to the British under the Treaty of Yandaboo in 1826, the court of directors appointed Dr. Wallich in 1827 to take stock of the inventories of the Burma forests. They were, he reported, unrivaled by any other sector of India, and the state ought to protect them. But a plan of exploitation and protection did not emerge. In 1829 timber merchants could still harvest teak on a royalty system without limit, provided the royalty would be paid to the government.

But soon the teak inventories ran low, and in 1841 the directors appointed Captain Tremenheere as superintendent of forests in Burma to regulate the harvest. Once again, as in 1823 with Captain Watson, opposition to the cancellation of timber contracts, leases, and the royalty system led the directors to overrule the superintendent and restore the former royalty system.

The development of forest policy, 1850 to 1870

In 1848 a new governor-general arrived in India. James Andrew Broun Ramsay, Earl of Dalhousie, became governor-general of India on the 12th of January 1848 and stayed until February 29, 1856. He set in motion the administrative and legal

---

[30] W. J. Hooker wrote *Exotic Flora: Containing Figures & Descriptions of New, Rare, & Otherwise Interesting Exotic Plants, Especially of such as are Deserving of being Cultivated in our Gardens* (Edinburgh, 1823–1827).

[31] Alexander Gibson to John Hooker, March 24, 1841, India Letters, Royal Botanic Gardens Archives, Kew, Surrey.

[32] Alexander Gibson, "A general sketch of the province of Guzerat, from Deesa to Daman," *Transactions of the Medical and Physical Society of Bombay* 1 (1838): 1–77.

[33] Stebbing, *Forests of India*, vol. I, 118.

structures that became the Indian Forest Department and the model of forestry for the empire. These eight years saw the settlement of administrative reforms for which posterity dubbed Dalhousie "the Great Proconsul."

Under his administration the crown and the East India Company annexed eight states: the Punjab, Pegu, Satara Samalpur, Jhansi, Nagpur, Berar, and Oudh. A utilitarian and "a curious compound of a despot and a radical," Dalhousie crossed the characteristics of a "Tory gentleman and the scientific Benthamite administrator."[34] To Dalhousie, government intervention held no ideological complications. As a member of the Board of Trade, he helped design the regulation of the British railways and saw anything upon which "the whole machinery of society will be stimulated" as fair game for intervention.[35]

In fact, he set in motion in India the state ownership and regulation not only of forests but also of railways, first with a publicly guaranteed stock company which constructed the railroads, and later with subsequent administrations carrying on his vision, complete nationalization of the rails.[36] Railways to Dalhousie were like forests, "a means of economic development," and should be regulated or owned by the state.[37]

Unquestionably Dalhousie was an unapologetic imperialist. When he reached Calcutta for his new appointment he proclaimed "we are Lords Paramount of India, and our policy is to acquire as direct a dominion over the territories in possession of the native princes, as we already hold over the other half of India."[38] By 1852, due to hostilities between British traders and officers of the king of Ava, Dalhousie annexed the province of Pegu, which contained a large reserve of teak. With the annexation of the Punjab and Oudh, the government acquired control over vast forest lands that required management and settlement.

To this question Dalhousie directed his energy. The eminent botanist and geologist J. D. Hooker, who met the new governor-general while sailing to India in 1847, advised Dalhousie. As director of the Royal Botanical Gardens at Kew, Hooker's advice carried weight with Dalhousie and proved pivotal in swinging him toward conservation.[39] Hooker alerted Dalhousie to the tragic economic and climatic effects of deforestation and recounted the example of St. Helena, which

[34] G. M. Young, *Victorian England*, quoted in Eric Stokes, *The English Utilitarians and India* (Oxford, 1959): 248, 257.

[35] ibid., 253.

[36] Daniel Thorner, "Great Britain and the Development of India's Railways," *Journal of Economic History* 11 (1951): 389, 393, 397; Daniel Thorner, *Investment in Empire: British Railway and Steam-Shipping Enterprise in India, 1825–1849* (Philadelphia, 1950), 46, 91–93, 119, 176; Gustav Cohn, *Zur Geschichte und Politik des Verkehrswesens* (Stuttgart, 1900), 91; "Administrative Report on the Railways in India," *Indian Forester* 18 (1892): 229.

[37] John A. Armstrong, *The European Administrative Elite* (Princeton, 1973), 287; Rondo E. Cameron, "The Credit Mobilier and the Economic Development of Europe," *Journal of Political Economy* 61 (1953): 465–478.

[38] *Cyclopaedia of India and of Eastern and Southern Asia*, 3rd edn. (Madras, 1857), s.v. "Dalhousie."

[39] W. B. Turrill, *Joseph Dalton Hooker: Botanist, Explorer and Administrator* (London, 1963), 50–51.

he had visited in 1842. Dalhousie responded to Hooker's advice with an enthusiasm that, together with the literature of Indian surgeon and writer Edward Green Balfour, transformed him into a forestry enthusiast.[40]

A further influence on Dalhousie proved to be his tour of the Punjab shortly after annexation. The Punjab, lying in the northwest corner of India and rich in unexploited timber resources, particularly in the hilly terrain of its western territory, had been eyed by the British well before Dalhousie. Lord Ellenborough, in his correspondence with the Duke of Wellington and Queen Victoria in 1843, predicted that the "time cannot be very distant when the Punjab will fall into our management and the question will be what we shall do as respects the hills."[41]

Dalhousie did not hesitate to take the opportunity of annexation when a suitable pretext arose. He argued in a letter to the secret committee in 1849 that the Sikh government violated the treaties of Lahore and Bhyrowal because back debts were owed to the British government. He also charged the Sikhs with instigating unending hostilities – a situation exacerbated by the minority of the young maharaja, who could not keep control of their subjects. In a letter to the home authorities he added that if the Punjab were annexed it would pay for itself, and the British Empire would gain "the historical jewel of the Mughal Emperors in the Crown of his own Sovereign."[42] His action in installing revenue-producing forest administration in the Punjab, the first establishment of the Indian Forest Department in India, is reason to believe that he thought the forests the worthiest jewel in the territory and a means of paying for the expansion of imperial rule.

But central to the work of setting up India's first forest department was the necessity of declaring all property of the former maharaja state property.[43] The proclamation of March 30, 1849, issued in Punjab, declared the territory to be in the British Empire. Dalhousie promised to rule with beneficence those who submitted to British authority and to punish those who violated the laws of the country. The new regime included, for the first time in India, a centrally organized and policed forest system.[44]

The people in the city of Amritsar in the Punjab received the proclamation with demonstrations of wild approval, which encouraged Dalhousie to add other territories. But the new annexations made the difficulty of financing the expanding empire in India more pressing. Financing the annexation of Pegu particularly

---

[40] Grove, *Green Imperialism*, 454.

[41] Edward Law Ellenborough, *History of the Indian Administration of Lord Ellenborough: in his Correspondence with the Duke of Wellington: to which is Prefixed, by Permission of Her Majesty, Lord Ellenborough's Letters to the Queen During that Period*, ed. Reginald Colchester (London, 1874), 399.

[42] *Parliamentary Papers*, The Punjab: 1847–1849: no. 53: 655–665; J. G. A. Baird, *Private Letters of the Marquees of Dalhousie* (London, 1911), 62.

[43] National Archives, New Delhi, *Secret Consultations*, April 28, 1849, no. 41.

[44] National Archives, New Delhi, *Secret Consultations*, April 28, 1849, no. 21; National Archives, *Governor General's Camp Letters to the Court of Directors*, April 20, 1849, no. 21.

6 Reserved pine forest in the Punjab, India. Early environmentalists believed that deforestation of hilly areas reduced stream flow and eroded soil.

concerned Dalhousie, since it constituted the furthest eastern reach of the British Indian Empire and defined the boundary to the Far East. The province of Pegu comprised the kingdom of Pegu and territories 50 miles beyond Prome. It was heavily forested around the Sittang River and its tributaries, as well as the Irrawaddy, comprising 32,250 square miles with almost 600,000 people. Dalhousie discussed the question with his advisors at length. The attempt would cost an enormous amount of treasure, and therefore he needed to obtain reparation for annexation costs and funds for the additional cost of administration and security.[45]

Establishing a forestry department was vital. Pegu boasted not only rich veins of coal but also oil fields and teak. Ownership of Pegu meant revenue, including revenue-producing forests that simultaneously deprived the court of Ava

[45] A. P. Phayre, *Report on the Administration of the Province of Pegu for 1854–1855 and 1855–1856* (Calcutta, 1857), paras. 2–12. National Archive, New Delhi, *Secret Consultations*, Dec. 29, 1852, no. 15; National Archives, New Delhi, *Letters to the Secret Committee*, Jan. 5, 1853, no. 3, paras. 2–4.

7 Reserved forest in the Gangetic delta, India. The "household of nature" preserved in this forest is clear.

of resources to continue hostilities.[46] The object proved too tempting to resist, and thus the need for timber and revenues fueled the extension of empire.[47] These considerations overruled a "regard to the natural formation of the country." The boundaries, the commissioners unanimously decided, would be fixed to include the rich southern forests and to extend no further than 6 miles north of the Myede.[48]

Control of this teak-rich portion of Burma gave command of the river Irrawaddy and, Dalhousie wrote to the king of Ava, "full control over the trade and the supplies upon which your kingdom so largely depends." He urged him to sign a treaty and cede the territory without further hostility.[49] After the conclusion of the treaty Dalhousie appointed a forest officer to collect revenue.[50] He also sent a mission to

[46] Regarding the expense of annexation, see National Archives, New Delhi, *Letters from the Secret Committee*, Sept. 6, 1852, no. 1524, para. 21. Regarding the revenue to be derived from Pegu, see National Archive, New Delhi, *Secret Consultations*, Nov. 26, 1852, no. 1.

[47] National Archives, New Delhi, *Secret Consultations*, Feb. 16, 1853, nos. 11 and 12; D. G. E. Hall, *The Dalhousie–Phayre Correspondence: 1852–1856* (London, 1932), letters nos. 8 (16–18), 15 (31–33), 19 (37–38).

[48] National Archives, New Delhi, *Secret Consultations*, April 29, 1853, nos. 71 and 73.

[49] National Archives, New Delhi, *Secret Consultations*, Dec. 29, 1852, no. 4.

[50] National Archives, New Delhi, *Secret Consultations*, Dec. 29, 1852, no. 15.

8 Transporting deodar sleepers for broad-gauge railway construction in Kashmir.

9 A rare photograph of a forest rest house in Burma, with a woman standing in the fore corner. R. S. Troup.

the court of Ava to cement the new relationship and take inventory of the territory as they went through the region. Oil fields, coal, forests, all were to be observed and placed into the imperial topography.[51] Cashing in on the booty, in 1855 the government imposed a duty upon teak.[52]

But the new governor-general wanted more than Pegu. When Colonel James Outram arrived as resident in Oudh in 1855 he reported to Dalhousie in a minute of June 18, 1855 that the government of Oudh abused its powers, oppressed its subjects, and engaged in no reforms or improvements of any kind, leaving the territory in a "most deplorable" state.[53] With the agreement of the court of directors, Dalhousie annulled the treaty of 1837 and then of 1801, and demanded the king of Oudh to hand civil and military control over to the British for the full annexation of the country into the empire. Since he had met no resistance from the directors about the Punjab, Pegu, Stataram, or Nagpur, he did not expect a dispute with them over the annexation of Oudh.[54]

The governor-general's prediction proved correct, and the court of directors agreed that Oudh ought to be absorbed.[55] They authorized him to assume "authoritatively the powers necessary for the permanent establishment of good government throughout the country," and this included the new establishment of forestry.[56] Dalhousie wrote to a friend in England that he had delivered to the queen not only 5 million more subjects but £1,300,000 more revenue and 25,000 square miles "without a drop of blood and almost without a murmur."[57] The new territory would be administered on the Punjab model, with the judicial commissioner overseeing the administration of civilian and criminal proceedings, leading to, it was hoped, the development of the country economically and legally. With the new department of forestry, he introduced some environmental protection as well.[58]

---

[51] Dr. John Forsyth had instructions to collect information on the climate, Captain Rennie for the course of the Irrawaddy. Major Grant Allan had instructions to collect information for military questions, Golesworthy Grant for the natural features of the land, Dr. Oldham, as superintendent of the Geological Survey of India, to ascertain the commercial value of the coal and oil fields. See E. P. Stebbing, "Pioneers of Indian Forestry; Captain Forsyth and the Highlands of Central India," *Indian Forester* 30 (1904): 339. Also National Archives, New Delhi, *Secret Consultations*, June 29, 1855, no. 2; National Archives, New Delhi, *Letter to the Secret Committee*, Aug. 8, 1855, no. 45.

[52] National Archives, New Delhi, *Political Letters to the Court of Directors*, Aug. 22, 1856, no. 83, paras. 10–18.

[53] National Archives, New Delhi, *Pol. Con.*, Dec. 28, 1855, no. 319, paras. 2, 4, 10–27, 28–44.

[54] Baird, *Letters*, 33, 169, 262.

[55] National Archives, New Delhi, *Political Letters from the Court of Directors and the Secretary of State*, Nov. 21, 1855, no. 33, para. 2.

[56] National Archives, New Delhi, *Political Letters from the Court of Directors and the Secretary of State*, Dec. 10, 1856, no. 47, para. 4.

[57] Baird, *Letters*, 369; National Archives, *Political Letters from the Court of Directors and the Secretary of State*, Dec. 10, 1856, no. 47, paras. 2–3.

[58] National Archives, New Delhi, Miscellaneous Branch, *Consultations*, May 27, 1859, nos. 381–382, paras. 145–159; National Archives, New Delhi, *Political Consultations*, June 6, 1856, no. 193, paras. 19–22, 47–56;

While the above annexations occurred, Dalhousie appointed a superintendent of forests in Pegu. The British government in India made it clear that "all the forests are the property of Government, and no general permission to cut timber therein will be granted to anyone."[59] The new superintendent, John McClelland, received instructions to "mark the trees which may be bought and felled." Accompanied by Captain Phayre, Commissioner of Pegu, he set out to discover a better system of management and report his suggestions to Dalhousie. His report dated April 5, 1854 included a description of the new forests in such detail and with such an eye for potential use that it lent credibility to his proposals.

He observed first that a very sparse population had settled in the forests, particularly in the hilly areas where the teak trees had not yet been exploited. Roads were practically nonexistent and dirt paths crisscrossed the territory by which carts sometimes passed. His first telling observation regarded the size of the teak trees. Teak trees on the lower elevations where extraction was easier were small, due to the removal of the large-sized trees. The teak in the hill forests appeared perfect, "growing on a grey stiff sandy clay in company with several species of large timber trees, which far outnumbered it in quantity." McClelland estimated that a "teak forest" meant one teak tree for 500 other trees of heterogeneous species. Teak was not diffuse, but "confined to certain localities of small extent where it constitutes the prevailing tree for a few hundred yards, seldom for a mile continuously."[60] Due to the method of extraction in the forest lowlands, the best teak grew high up in the forests where, if the streams (*coungs*) were exploited correctly, the timber could be floated down in the rainy season thus saving the construction of a large network of roads. McClelland saw the glaring need for scientific forestry when he observed that young trees of only 2 to 4 feet in girth were harvested in the lowlands, destroying the process of natural seedings that an older tree provided. This waste cut the future yield significantly to satisfy the urge for an immediate profit with no concern for either the forest or for future revenue.[61]

The structure of marketing also wasted trees. Merchant monopolies paid the government little and charged customers dearly. Further, the natives in the south Pegu forests received advances from merchants in Rangoon, who paid them to float timber downstream to market. The men did not care if the trees were hardly more than saplings. He observed that "In Pegu there is no such class of foresters or professional woodcutters, that is, persons who have been accustomed to earn their bread by forest work, or who can be thrown out of accustomed employment or be in any way injuriously affected by any alterations in the forest laws or rules." This led to a state of lawless exploitation and the depletion of the forests for posterity.[62]

McClelland proposed two simple rules to rectify the situation: (1) a single duty on logs – not a percentage of their worth – to discourage the cutting of undersized trees; and (2) the reservation of forests by the government, where

---

[59] Stebbing, *Forests of India*, vol. I, 244.   [60] ibid., 247.   [61] ibid., 248.   [62] ibid., 249.

merchants could only extract timber marked by foresters. To enforce these rules he recommended that revenue stations be established at the riverbanks of all the rivers below the teak forests. There logs could be inspected for the official markings and a duty calculated. Purchasers of timber would receive information from the forest department, who would then advise the client and invite a tender fair to both parties. Those who felt they had claims on existing timber stands ought to be given so many days to register their claims for a settlement with the government and, after that point, no disputes regarding the ownership of the new timber stands ought to be considered.[63] He concluded with a sentiment which would become universal in the establishment of forestry reserves for the rest of the colonies and the United States:

a forest may be regarded as a growing capital, the resources of which are the young trees, and unless these are preserved and guarded to maturity, it is obvious the forest must necessarily degenerate from the nature of an improving capital to that of a sinking fund, which, within a given time, must become expended. The loss occasioned by the removal of an undersized tree is not merely the difference of value as compared with a full-grown tree as a piece of timber, but must be estimated by the number of years the forest may be deprived, by its removal, of the annual distribution of its seeds, and the time it would otherwise have taken to arrive at maturity.[64]

McClelland also concluded that the planting of new teak forests would never replace conservation, for "if we fail in the comparatively simple duty of preserving the old forests, we can scarcely hope to succeed in the more difficult task of creating new ones." A forest ought not to be considered only for a single species, because "an exclusive search and use of teak alone . . . [has] caused other descriptions of timber to be entirely overlooked . . . Time and necessity will in due course render these and other resources of the forest better known, [such] as oils, gums and textile material."[65]

But McClelland's report did not go to Lord Dalhousie unaccompanied. Captain Phayre, the commissioner of Pegu, disagreed sharply with McClelland, and in a letter accompanying the report to the Secretary he denounced the suggested charge of 8 rupees per log as exorbitant. He recommended that the royalty be no higher than the royalty in the Tenasserim forests, which had been so devastated. He considered "the principle of Dr. McClelland's plan, that, namely of preserving the forests by means of rates of duty on timber so high as to render the working of them profitless, or barely remunerative, as being essentially wrong: I could not, therefore, in any case recommend its adoption." He then informed the government that it ought to implement a 4-rupee a log duty, half McClelland's recommendation. He also criticized McClelland for making reference to planting new teak trees in the district.[66]

---

[63] "On Forest Settlement and Administration," *Indian Forester* 19 (1893): 24.

[64] Stebbing, *Forests of India*, vol. I, 251.   [65] ibid., 252, 255.

[66] Captain Phayre, commissioner of Pegu, letter dated July 24, 1854, "To the Government of India," as quoted in Stebbing, *History*, vol. I, 256.

Stebbing mused that perhaps the "rather careless manner" in which Captain Phayre had written his letter drew Dalhousie's attention – in a way unfortunate for Captain Phayre and crucial to the development of forestry policy in India. Dalhousie seemed piqued and annoyed at Phayre's lack of detail and the cavalier criticism of McClelland's carefully written and carefully argued report. Mr. C. Beadon, Secretary to the Government of India, answered for Dalhousie and communicated to Captain Phayre Dalhousie's outlines for the future Forest Charter, which proved so critical to the development of modern environmentalism.[67]

Beadon stated icily that His Honor in Council (Dalhousie) "found it somewhat difficult to understand exactly what it is proposed to do." Phayre seemed not to have "seen the full force of the principle which was laid down by the Government at the annexation" and that "the teak timber should be retained as State property." Therefore, the suggested duty of the logs at 4 rupees did not include a price for the log, since a duty is not a price but a tax, and if the duty is 4 rupees the state is getting essentially no price for the wood. "You reduce the rates proposed by Dr. McClelland by one half, on the ground that at such high rates, the forests could not be worked at a profit, and that such charges on timber would destroy Rangoon as a shipbuilding port. You, however, adduce no proof in support of this assertion."

Dalhousie then argued, through Beadon's letter, that he "cannot think that Dr. McClelland's rate of duty would have that effect" and approved the idea of a timber auction in which the government would get its price for timber at whatever the market would bear, over the price of the duty. This timber tax still guaranteed the framework of market competition and the exclusion of monopoly, while bringing to the government a solid revenue. In addition, Dalhousie liked the prohibition of merchants from the forests and the construction of revenue stations on the rivers for effective collections. He also agreed with McClelland that the planting of new trees over the conservation of the great tracts of natural woodland made little sense. From this report came the memorandum of the Government of India, August 3, 1855, which Stebbing dubbed the "Charter of Indian Forestry."

This charter, based upon the restriction of private interests contained in McClelland's report and Dalhousie's response, pointed out that annexation was a ruling principle, and that forests must be considered by definition, if not privately owned, then not wastelands but state property. This ringing bill of rights for the government had various consequences. Not only did private interests not have the right to extract timber free with a set royalty, but had no claim to standing dead trees, felled trees, or green trees – all belonged to the state, along with profits from the sale. In this respect Lord Dalhousie's pronouncement proved precedent-setting. For with the establishment of forest areas as absolute state property, the charter required proper management of the forest areas, and this meant scientific

[67] *Parliamentary Papers*, Dalhousie's "Minute on Forest Policy," Aug. 3, 1855, Fort William; also see Stebbing, *Forests of India*, vol. I, 257.

forestry. From this new legal definition of the forest as state property (and later of nature as state property) grew the policies and practices of environmentalism.[68]

Innovation and stabilization: 1855 to 1900

The stabilization of forestry in this period was due to the utilization of German-trained scientists and foresters, and in particular one, Dietrich Brandis. In 1866 Brandis secured the services of Schlich and Ribbentrop, both German-trained. Schlich had served his probation in Burma and in 1870 had been transferred to the Sind. Ribbentrop began his work in the Punjab. Schlich and then Ribbentrop followed Brandis as inspector general of forests. Schlich served from October 1881 to December 1888 (however he went home in February 1885 to found the forestry department at Cooper's Hill) while Ribbentrop served from 1884 to 1889.[69]

Stebbing argued that while the German methods of forestry were valuable, Germany's "hard and fast methods" based on "axiomatic dicta and calculations" did not transplant well to India. Though at first foresters copied European methods slavishly, they soon learned to adapt their training to the varied Indian circumstances, where foresters required less dogma and greater individualism and flexibility to manage forests as large as some European countries.[70]

But before large areas could be "settled," legislation had to precede the settlement, which would deal with the right of the user to collect produce. The need to settle claims against forest areas led to the rise of the "multiple use" doctrine that later served as the model for empire forestry as a whole. In some cases the rights of the user were acquired by *sanad* (grant), while for communal use the state had settled for the trees only and given the right of the soil (grazing primarily) to the villagers. Further, there were islands of private land that required access through state lands, or in which the property was private or communal but where the state had rights over certain trees, such as the royal teak trees of Burma. Finally, forests were settled to give absolute rights to the state. It is this last category that constituted the primary innovation of Indian forestry. A legal precedent was put in place

---

[68] Scientific forestry was defined as forestry methods based on experiment and verification of principles. See "Scientific Forestry," *Indian Forester* 33 (1907): 89.

[69] J. S. Gamble, "Instruction in Forestry at Cooper's Hill," *Indian Forester* 18 (1892): 292. The list of Inspectors General of Forests is as follows: Captain E. C. S. Williams, RE from April 13, 1865 to May 7, 1866; Dr. H. Cleghorn from May 7, 1866, to March 14, 1867; Colonel G. F. Pearson, MSC from Jan. 29, 1871 to Dec. 20, 1872; Mr. B. H. Baden Powell, ICS, CIE from Dec. 30, 1872 to April 8, 1874; Colonel F. Bailey, RE, LLD from Aug. 3, 1887 to Oct. 31, 1887; and Mr. H. C. Hill, CIE from Aug. 7, 1889 to March 1, 1894, from Feb. 19, 1895 to April 1, 1896, and from July 8, 1899 to Oct. 8, 1899. See Stebbing, *Forests of India*, vol. II, 292.

[70] G. F. Pearson, "The Teaching of Forestry," *JSA* 30 (1881–1882): 422–428; F. J. Bramwell and H. Trueman Wood, "Education in Forestry," *JSA* 30 (1882): 879; J. S. Gamble, "The Advantage of Preliminary Practical Work in the Training of Forest Officers," *Indian Forester* 18 (1892): 96; M. R. K. Jerram, *Report on the Group System of Natural Regeneration in Germany and its Application to Indian Forests* (Simla, 1913).

for the state to acquire property rights as absolute as private property over large areas of previously unsettled land.[71]

This legal definition involved clarifying the distinction between the rights and claims of individuals from the state. The first legal act to do so was the Forest Act (Act 7) framed in 1865. It had been assumed that the force of law would be sufficient to enforce the act and clarify use right. But a multitude of rules were required to make the separation of rights work, and so in 1875 the Hazara Rules were passed. These codified the forest department practice into law, with the rules recast and amended in 1878 and then again in 1893, clarifying the forests as absolute state property under the Land Revenue Settlement.[72]

Regional acts were also passed to create state forests out of the particular circumstances of the territories. A Revised Bill and Memorandum Act of 1865 was submitted by Brandis and passed by the government of India in 1868. In it, the local governments were to submit opinions for forest rules in their respective areas by 1871. The forest conference of 1873–1874 held at Allahabad pointed out the deficiencies of the 1865 bill and recommended changes that were adopted in Act 7 of 1878, which extended to all the provinces of India (except Coorg, Madras, Burma, Berar, Ajmer and Baluchistan, and Hazara in Punjab).

This act clarified the words "reserved" and "protected," making clearer distinctions between them (see chapter 4 below). The first draft of the act (1865, Act 7) made clear the process by which the state acquired absolute property rights over a forest by mandating that the government hold hearings into and check recordings of the rights of individuals over the given land areas. It also gave the state authority to decide if the absolute property designation of "reserved" was feasible. If so, then the state would proceed to buy out the existing rights from individuals with any claims. If the inquiry showed that the reservation of forest lands was not possible, then the state would move to protect the forest with a shared multiple-use designation.[73]

Since foresters felt that "protected" status did not preserve the whole economy of nature, they pushed for "reserved" wherever possible, so that by 1889–1890 there were 50,000 square miles of reserves and 20,000 square miles of protected forest. By 1899 there were 81,400 square miles of reserves and only 8,800 miles of protected. The smaller number of protected forests meant that settlements of rights continued and the government bought out user privileges and converted recently designated protected forests into the higher category of reserved.[74]

---

[71] "Grazing and Commutation in the C.P.," *Indian Forester* 18 (1892): 415; "Grazing in Forest Lands," *Indian Forester* 26 (1900): 235; "The Effects of Grazing on Forests," *Indian Forester* 26 (1900): 283.

[72] "The New Draft Rules Regarding Settlement and the Positions of Revenue and Forest Officers," *Indian Forester* 19 (1893): 108.

[73] Gem, "A Plea for Protected Forests," *Indian Forester* 19 (1893): 123–136.

[74] "On Forest Settlement and Administration," 24. BHBP, "Protected Forests," *Indian Forester* (1893): 294. Gem, "Protected Forests," 57. Burma required special legislative acts due to the size of its

The quest for imperial power and revenue does not explain fully the radical forestry program initiated in India.[75] Though most forests produced revenue, an examination of working plans from the 1860s to the 1920s reveals an unexpected tolerance for unprofitable forests. The plans for the Mudamalai, Kumbarakolli, and Benne forests in the Nilgir district (Madras Presidency) exemplify this. The Indian forest service leased the forest from an Indian merchant after a series of lumber cuts left the forest with little teak, severe soil erosion, and prodigious patches of elephant grass that made fire protection almost impossible. As part of the leasing agreement, the Indian forest service insisted that merchants leave hill tribes undisturbed, and allow them to live off the forest in the traditional manner, free of charge.[76] But after forty-three years of management, from 1862 to 1905, the working circle produced a running *deficit* of over 10 percent of operating costs. Again the Indian forest service decided, in 1907, to keep the forests under management for another forty years of revenue loss. Then in 1909 officials proposed to run the forests for up to 100 years of projected loss. This, even though the humid condition exacted a terrifying toll on the health of foresters. One official lamented that "ever since the forest has been in the charge of the Forest Department, the Mudumalai ranger has not been in good health; ranger after ranger has broken down and the cadre for the circle is so limited that men have to be kept on duty as long as they can crawl about and sign their names."[77]

Why keep an unhealthy and unprofitable forest? Particularly when "the net cost of management will be considerable" and no profit could be realized "in from 90–100 years."[78] Why would inspectors "proposes [*sic*] entirely to subordinate financial working to sylvicultural workings?"[79] Because, as the working plans

---

forests and the limitation of its forest staff. The Presidency of Madras united with the greater Forest Department under the Forest Act 1882 and the Burma Forest Act 1886. As Stebbing pointed out, "It had taken many years, but we had thoroughly learnt our lesson by 1887 in India. In other parts of the Empire this lesson, however, still remained to be assimilated by the Administrations responsible." *Forests of India*, vol. II, 470.

[75] The German model of forestry, utilized for technical expertise, did not merely transplant to India. For instance the influential German silviculturalist Faustmann developed a "steady state" model of forestry that maximized profit by a system of age rotation that harvested stock when its rate of growth matched the real interest rate. Indian foresters did not use this model.

[76] *The Control Journal of the Mudumalai Leased Forests and the Benne Reserved Forest* (Madras Forest College, 1924): 2, 7–8, 15. This journal contains the accounts of forest managers over a sixty-year period, edited, and in parts summarized, by J. H. Longrigg, principle of the Madras Forest College. The *Control Journal* is privately held by the Longrigg family. The working plan explains that the "Detected offenses are few in these forests. The occupiers of land within the forests are given timber, bamboos, and grazing free of cost; the only offenses which are serious although seldom detected, are those of incendiarism, willful or accidental" (see page 52). For a fascinating account of foresters rescuing indigenous jungle people from slavery by Indian merchants, and for a rich collection of personal remembrances by forest rangers and their families, see Mary McDonald Ledzion, *Forest Families* (London, 1991).

[77] *Control Journal*, 18.   [78] ibid., 13.   [79] ibid., 17.

argue, (1) shortcuts do not lead to forest restoration,[80] (2) forests protect water flow and "are of importance climatically,"[81] (3) the rapidly reduced forest areas of the world "is [*sic*] a matter of world wide importance,"[82] and (4) the forest would otherwise be "handed back to its owner to be cut to ribbons soon after by timber contractors" and the restoration of the soil, and the humus, essential for the household of nature, would be lost.[83] To allow that would be a "discredit to government forest administration."[84] Clearly ecological concerns were primary to the conservation ethic.

The series of legislation that followed the 1865 Indian Forest Act built and extended the original Forest Charter sponsored by Dalhousie. An act of 1927 provided for the further protection of wildlife in India; this was amended in 1930, 1933, and 1948 to allow the central government and the state government to declare wastelands and forest areas not privately owned a reserved forest. These new reserved forests could prohibit not only the removal of trees and minor forest product, but also hunting, fishing, and trapping.

This status, first envisioned by Brandis and actively promoted by Ribbentrop and Stebbing, is, outside of a park or wildlife sanctuary, the highest form of protection. Next, and lower in the level of protection, comes a village forest, in which a village community is given the right to harvest timber and graze and gather produce under government rules and direction.

When Stebbing looked back over the progress of forestry in the forty-five years from 1855 to 1900, he expressed amazement at the accomplishment: despite major obstacles, by 1900 8 percent of the entire land area of greater India had been placed under the protected control of the forest department. The innovative methods behind this monumental achievement, discussed in the following chapter, captured the imagination in this period, not just of other foresters or would-be foresters, but also of a general public throughout the colonies and much of the world.

[80] ibid., 8: "Proceedings of the chief conservator of forests, proceedings no. 23, 19 January 1923," appendix.
[81] ibid., 13.   [82] ibid., 40.   [83] ibid., 41, 110.   [84] ibid., 40.

4

Environmental innovation in British India

The creation of the Indian Forest Department

Two years after the Forest Charter came the Indian Mutiny, which burst over colonial India in May 1857 and proved the catalyst for both greater deforestation and a militant environmental response. Dalhousie had left India with fair prospects for the colony, declaring that India had been "in peace without and within" with "no quarter from which a formidable war could reasonably be expected at present."[1] Then quite unexpectedly Indian soldiers rebelled at bullets greased with animal fat. In the colossal effort to put the mutiny down, the British found themselves faced with a dearth of rapid communication – roads, railways, and canals. After the rebellion, railway construction became the priority of Lord Canning, who succeeded Lord Dalhousie as governor-general in 1856.

At the time of the mutiny the beginnings of environmental forestry had been undertaken in Bombay, Madras, and Burma, with Dietrich Brandis setting up a scientific forest administration under the authority of the charter. Also notable for having formed new forestry departments were the Punjab, Oudh, Bengal, and Assam.[2] But when the East India Company found its authority revoked after the mutiny and the double government of crown and company ended, a far more centralized regime emerged, with profound implications for the environment.[3]

In 1864 the governor-general appointed Dietrich Brandis as the first inspector general. Concurrently he also established the Indian Forest Department as an organized state department under the umbrella of the Indian exchequer. In 1871 the Department of Revenue and Agriculture, newly established, oversaw the forest department, itself under the umbrella of the home department.[4]

---

[1] Col. Pearson, "Recollection of the Early Days of the Indian Forest Department, 1858–1864," *Indian Forester* 29 (1903): 313–319.

[2] Tuscan, "Forest Administration in the Central Provinces," *Indian Forester* 19 (1893): 45, 332.

[3] Pearson, "Recollection of the Early Days," 313–319.

[4] In 1879 the Department of Revenue and Agriculture had been abolished, but it was reformed in 1881 with the forest department again under its umbrella. Throughout the period of this investigation the home department held the final responsibility for the forest department. There is no reliable history

Dietrich Brandis held the position of Inspector General of Forests between 1864 and 1883, followed by Wilhelm Schlich from 1881 to 1888, Berthold Ribbentrop from 1884 to 1899, and E. P. Stebbing from 1900 to 1917.[5] Throughout the tenure of these inspectors general, conservators were appointed to Indian provinces in the following order. Bombay, Madras, and Burma were placed under a conservator at the appointment of Brandis (Bombay 1847, Madras 1856, United Burma Provinces 1857). Then were added the North-Western Provinces in 1860, the Central Provinces also in 1860, Oudh in 1861, the Punjab in 1864, Coorg in 1864, Bengal also in 1864, Assam in 1868, and Berar in 1868.[6] By the end of 1868 the forest department had administrators in every province of the subcontinent, so that by 1885 the inspector general oversaw 10 conservators of forests, who in turn oversaw 55 deputy conservators of forests, 38 assistant conservators of forests, and thousands of forest guards.[7]

The new forest department did not have specially trained forest officers to accomplish the work, and the appointments were filled by men from other branches of government service, often those who in some way showed themselves fit for "forest life," sometimes naturalists, military men, or sportsmen. Some personnel were gained by merging previous agencies with the department. For instance, in the Madras Presidency the Jungle Conservancy Department amalgamated with the forest department.[8] Recruitment also depended upon "young gentlemen," mostly from Britain, who until 1891 were awarded jobs in local government on a system of patronage given "to young men of European extraction . . . [who] grasped the situation more rapidly, and came into the service in considerable numbers."[9] Alongside the young gentlemen were "many a useless Baboo" who usually did not rise as quickly as the English men, due, Ribbentrop claimed, to a lack of interest in forest work.

This rather loose personnel policy changed with the reorganization of the department in 1891 and with the instigation of the forest schools. The reorganization separated the forest department into two grades, the imperial service and the provincial service, both working for the Indian Forest Service.[10] The provincial service

of the formation of the Indian Forest Department or of forestry in nineteenth-century India. Two titles intended to address the issue of forestry in India do not meet western standards of scholarship. The first is Ajay Rawat, ed., *History of Forestry in India* (New Delhi, 1991). This work addresses few issues of forestry in the nineteenth century and contains collected lists of indigestible data, often without identifiable argument. The second, G. S. Padhi's *Forestry in India* (Dehra Dun, 1982), provides an eclectic assortment of data on forestry subjects, but also without an identifiable narrative scheme or argument. Neither work is helpful to a student of forestry or environmental history.

[5] During the tenure of these Inspectors-General numerous individuals officiated when Brandis, Schlich, Ribbentrop, and Stebbing were out of India. See ch. 3 note 69 for full details.

[6] Tuscan, "Forest Administration in the Central Provinces," 45, 312.

[7] Ribbentrop, *Forestry*, 78, 80.   [8] ibid., 81, 82.   [9] ibid., 87.

[10] "Rules to Regulate Appointments and Promotions in the Provincial Forest Service," *JSG* 13 (1892): 101, 108; C. G. R. "Recruitment of Officers for the Indian Forest Service," *Indian Forester* 21 (1895): 450.

tended to attract native foresters graduating from the newly established forestry school in India (Dehra Dun), and the imperial service attracted Englishmen, usually with some college training. Personnel grew from 107 in 1885 to 10,508 in 1899.[11]

In 1864 the British had mapped and demarcated very little of India under a forestry organization, with the exceptions of limited areas in the Bombay and Madras presidencies, and in Burma. The extension of forest management to other districts and the development of an effective forest organization came together between 1864 and 1870. Timber, required in large amounts for export, and for sleepers and fuel by the burgeoning rail system – the largest in Asia – fueled the drive for further organization of annexed areas.[12] The process of reconstruction in the aftermath of the mutiny, initiated by Lord Canning and continued under Lord Elgin and then Lord Lawrence, attempted to fulfill the queen's proclamation "to stimulate the peaceful industry of India, to promote works of public utility and improvement, and to administer the government for the benefit of Her Majesty's subjects." This proclamation provided an administrative ideal and sanctioned further organization of forestry in the provincial and local administrations.[13]

The shift from the *laissez-faire* ethic to government intervention had been aided by the Orissa famine. When the monsoon rains failed in 1865, Orissa, then a division of Bengal and extending along the coast from Calcutta to Madras, experienced a severe drought and loss of crops. While Lord Harris, governor of Madras, had the benefit of developed roads and canals by which the market regulated prices and allowed food imports, the jurisdiction under the Bengal government lacked such infrastructure. Yet it still preferred to allow the market to regulate prices and handle the distribution of food. The result in Orissa was disastrous, with nearly a million people dead.[14]

The failure of the Bengal government pushed Sir John Lawrence to accelerate road construction, railroads, and canals. It also damaged the credibility of those who argued that the market alone should decide administrative practice. The mutiny and then the Orissa famine made a series of governors see noninterference as suicidal and inhumane. These catastrophes gave the forest department a powerful ideological boost and it found state intervention in forest areas was better received than had previously been the case.[15]

[11] C. W. A. Bruce, "The Reorganization of the Imperial and Provincial Services," *Indian Forester* 21 (1895): 465; C. G. R., "Recruits for the Upper Controlling Staff of the Forest Department," *Indian Forester* 20 (1894): 406; "The Future Training of the Upper Controlling Staff of the Service," *Indian Forester* 31 (1905): 361.

[12] "The History of a Railway Sleeper," *Indian Forester* 18 (1892): 446.

[13] Pearson, "Recollections of the Early Days," 313–319.

[14] Other famines in India, less severe, were linked to deforestation as well. See "Forests and Famine in Bombay," *Indian Forester* 27 (1901): 222; W. Tayler, "Famines in India: their Remedy and Prevention," *East India Association* 7 (1873): 143–182.

[15] "Forest Administration and Revenue Making," *Indian Forester* 31 (1905): 246.

Dietrich Brandis and Hugh Cleghorn set the stamp on the new forest organization in this period. Cleghorn had provided the model of forest conservancy in Mysore and Madras that Brandis emulated in Burma and eventually the rest of India, and had written the report for the British Association in 1850 that led in part to the Dalhousie legislative action in 1856. The government memorialized Cleghorn as a founder of Indian forestry in 1865, asserting that "His long services from the first organization of forest management in Madras . . . and in the Punjab . . . prepared the way for the establishment of an efficient system of conservancy and working the forests of that Province."[16]

Observing the need to set the new organization on a systematic and organized footing, Brandis and Cleghorn together drafted the Indian Forest Act 7 of 1865 for the consideration of the Indian government. It stands as the first India-wide forest legislation and the first broad-based environmental law to be applied in the nineteenth century. The newly annexed land, declared as government property, fell under the authority of the new act and determined the fate of an entire "household of nature" that included forests, soil, water quality, and pollution. The outlines of the legislation are paraphrased as follows, to give an idea of the environmental scope of the initiative.

(1) All trees, shrubs, and plants within the government forests were to be preserved. If specifically designated for exploitation, then only those selected by the department could be harvested. "All trees, shrub growth, and plants within Government Forests" were prohibited from being marketed, girdled, felled, or lopped; fires that would endanger all trees, shrubs, and plants were prohibited. The collecting and removing of leaves, fruits, grass, wood oil, resin, wax, honey, elephants' tusks, horns, skins and hides, stones, lime, or any natural produce of the forests was prohibited. Ingress into and passage through such forests was strictly prohibited without travel on an authorized road or path. Burning wood for lime or charcoal and the grazing of cattle within such forests were also prohibited.[17]

(2) All streams and canals in greater India came under regulation. Prohibited would be the closing or blocking up of streams or canals used for the transport of timber or forest produce and the poisoning of streams and waters in government forests used from timber processing. Stations on streams and canals would be built to regulate transportation on any river flowing through or from government forests; they would levy dues and revenues.

(3) "Regulation of the manner in which timber, being produce of Government Forests, shall be felled or converted."

---

[16] See Dietrich Brandis, "Pioneers of Indian Forestry; Dr. Hugh Cleghorn's Services to Indian Forestry," *Indian Forester* 31 (1905): 227–234.

[17] "Punishments for Cattle Trespass," *Indian Forester* 20 (1894): 410. Gathering charcoal for iron smelting was an additional strain on the forests. See "Charcoal for Iron Smelting," *Indian Forester* 19 (1893): 339.

(4) Regulation of government officers charged with conservancy of the forests and the collection of dues and revenues that were in place.
(5) Powers were given to the local governments to arrest violators, seize unlawfully collected forest property, prosecute, fine, and imprison malefactors.[18]

The rules of this act described the procedure of creating "reserved forests" and "unreserved forests," the method and legality of demarcation, and the behavior of visitors to preserved areas. From this point on all activities in the forest would require a license from the staff of forest offices. Officers were encouraged to apply the rules leniently at first, in order to wean jungle tribes slowly from communal practices of forest use to the new environmental policies.[19]

The proposed forest act drew rabid opposition from some quarters. In Madras the majority of tax collectors opposed the introduction of the act into the Madras Presidency and, supported by the Board of Revenue, issued a statement from their proceedings on April 16, 1868. The act, they argued, was "Legislation of a special nature." The act abridged "private right of every kind" and threatened property rights. For who, the collectors argued, is to say that all the annexed land certainly is governmental? "All the jungles and forests of this Presidency are within village boundaries" and villagers have "from time immemorial" the right to collect firewood and other products, such as manure and grass for grazing.

Now a new level of bureaucracy had been created, with forest officers to charge individuals before a magistrate, which "opens a wide and dangerous field for oppression and extortion" because even a "peon" could arrest and charge suspected violators. Thus the pettiest official could "harass the people at his own pleasure" and the people would be placed at the mercy of "uncontrolled officials." The preservation of nature would not be a forest officer's only motive for prosecution.[20]

The collectors also argued that "the so-called 'forests' of the Madras Presidency are, in truth, merely jungles with villages scattered through them," and to declare them off limits would require unprecedented policing and raise wild opposition from villagers. Since such a short period had elapsed since the government "began to exercise its forest rights," it would be best to merely mark out forest reserves carefully and maintain them with a minimum of interference. Otherwise enforcement would be monumental.

The collectors ended with the following statement that summed up the property rights nature of the debate. "It cannot be that such property is beyond the pale of the law, but if the doubt be as to Government being really the proprietor of any forest, the Act will not solve the difficulty, for Government cannot declare that to be their property which is not clearly their own."[21] Conservancy was valid, collectors concluded, only where it preserved the wooded areas that protect the water flow of a stream, and no more.

Brandis responded to the proceedings by arguing in his memoranda of March 23, 1869 that

---

[18] Stebbing, *Forests of India*, vol. II, 12.   [19] ibid., 13.   [20] ibid., 14–17.   [21] ibid., 17.

In the matter of forest legislation, I would draw attention to the peculiar difficulties under which a portion, at least, of the Bombay Forests are placed in the matter of protection. I understand that Government forests are frequently so interlaced with private forest lands that protection is impossible without a system of strict control over all timber, wood, and forest produce.

The solution to the problem raised by the collectors would be more, not less, control, Brandis argued.[22]

The collectors' recommendations were not adopted. One of the arguments that Brandis and Cleghorn made concerned money and security: strict regulation of forests created income and built up the communications of the country, which after the mutiny had become a necessity. Surplus revenue from the forest department was the "necessary consequence of conservancy management . . . in accordance with a rational and well-devised plan of operations."[23]

Thus Brandis made the argument that any expansion of governmental forests could be paid for by the revenue collected in any one year of their management. The money thus raised would become available to build access roads and canals for floating timber, simultaneously serving the purpose of building up the communications in India for the maintenance of colonial rule as well as for the stimulation of industry. Though the extraordinary expenditure of any one district may exceed its revenue due to the expense of canals or roads, it would "in the end prove directly remunerative."[24] Because the expansion of the forest department would serve so many ends, Stebbing later pointed out, "The secretary of State gave his cordial approval to these orders," and the orders were "sufficiently remarkable to be worthy of careful consideration in other parts of the British Empire."[25]

When Stebbing looked back over the progress of forestry in the forty-five years from 1855 to 1900, he expressed amazement at the accomplishment: despite major obstacles, by 1900 8 percent of the entire land area of greater India had been placed under the protected control of the forest department. The innovative methods behind this monumental achievement, discussed below, captured the imagination not just of other foresters or would-be foresters, but also of a general public throughout the colonies and much of the world.

Education and Indianization

Independence loomed as an issue in the twentieth century. The larger framework of the "Indian Problem" shook the Forest Department and involved it in larger constitutional questions. Most Indian forest officers of British or European extraction held views similar to the British position as a whole – expressed lucidly in

---

[22] ibid., 18.   [23] ibid., 20.

[24] National Archives, New Delhi, *Home Public Consultations*, John Lawrence to the Secretary of State, letter to Revenue-Forests, November 23, 1867, no. 24.

[25] Stebbing, *Forests of India*, vol. II, 26.

a report of 1934 on Indian constitutional reform. While not claiming infallibility or freedom from error, the British considered that they delivered to India its first stable government, shielded it from foreign invasion, maintained tranquillity, established the rule of law, a just administration, and an upright judiciary. To this list the forester would add the saving of nature.[26]

This optimistic view of British suzerainty did not capture the minds of all His Majesty's subjects in India. For decades the forest department wrestled with the complaints of Indians who felt excluded from higher posts. Officials agonized over this conflict until Parliament passed an act in 1919 that resolved the issue for the department. Before the act, while some Indians were highly placed, Europeans held a monopoly of higher posts. After the act, a conscious "Indianization" policy attempted to integrate and finally displace Europeans from the management of the department in preparation for eventual Indian self-rule or full independence.[27]

As early as 1878 the government of India had established a forest school at Dehra Dun specifically to train Indians for work in the service, particularly as forest rangers. Forest rangers had responsibility for tracts of land averaging 20,000 acres, with multiple forest guards. The school also qualified Indians to work their way up to subconservators. A resolution dated February 3, 1879 announced two classes of students for this new school: (1) forest rangers of native descent; and (2) graduates of the Thomason Civil Engineering College at Rurki, which trained Indians.[28] Classes in Hindi were substituted in 1884 to overcome the difficulty some students had with English, and admission expanded to include those Indians who had passed the entrance examination of an Indian university in English.[29]

Forest research in India itself began with the inauguration of the Dehra Dun Forest School. The Survey of India oversaw the school until the government of India took over administration in 1884. Later, in 1906, the school expanded under the direction of Sir Eardley Wilmot, Inspector General of Forests, into the Imperial Forest Research Institute and Colleges. When the forest service dropped the requirement that a degree from Oxford, Cambridge, or Edinburgh was required to serve in the upper echelons, the school began training forest officers in 1926.[30]

---

[26] Session 1933–1934, *Joint Committee on Indian Constitutional Reform*, vol. ɪ (part 1) (London, 1934), 3.

[27] Between 1919 and 1935 the following paragraph, taken from the preamble to the act of 1919, guided departmental personnel policy and "Indianization": "it is the declared policy of Parliament to provide for the increasing association of Indians in every branch of Indian administration, and for the gradual development of self-governing institutions, with a view to the progressive realization of responsible government in British India as an integral part of the empire" (ibid., 6).

[28] Proceedings of the Sub-Committee, *Public Service Commission* (Simla, 1888), 6.

[29] ibid., 7. For more information on the education of Indian students in the early part of the twentieth century, see Office of the High Commissioner for India, *Report of the Indian Students' Department, 1922–23* (London, 1923).

[30] *Explanatory Memoranda for the Central Board of Forestry, 7, 8, 9 May 1951, Forest Research Institute and Colleges* (Dehra Dun, 1951), 18, 19.

A difference of opinion about the employment of Indians in the position of conservator often split the Secretary of State and forest department officials. In 1878 a dispatch expressed the anxiety of the Secretary of State over employing natives in the higher echelons and demanded to know why non-British Europeans were promoted over natives. His anxiety stemmed from the need to satisfy growing Indian restlessness over foreign rule. Ever since the rebellion of 1857 many of the British in India keenly felt the dangers of alien rule. But not all higher level forest officers shared this political sensitivity.

In 1884 a group of Eurasian and Anglo-Indians petitioned the Secretary of State for more open admission of Indians into the government of India. The forest department forwarded the memorial to the Secretary of State and attached a report, which stated "anything like a general recruitment [of Indians] of the Staff in India would at the present state be fatal to the efficiency of the Department."[31] The Secretary of State agreed, reluctantly, but only because recruitment for the superior staff "must depend to a great extent on the facilities afforded for technical education." The school of forestry at Dehra Dun did not yet offer, he noted, the training necessary at that time.[32] A quick summary of the opinions expressed before a subcommittee of the Public Service Commission in 1888 illustrate these contrasting views.

Mr. C. Bagshawe from the North-Western Provinces stated that he saw no problem with natives serving as conservators so long as the training was sufficient, and would not close the door to promotion based on race. But a W. R. Fisher, Director of the Dehra Dun Forest School, complained that natives "dislike employment in the jungle, and prefer Judicial duties or Police work to exposure in the open air throughout the day." He also feared natives would delegate too much to their subordinates. E. E. Fernandez added that the educated native was impatient of the hardships that "other races would submit to with cheerfulness," and was also "influenced by caste prejudices." Generally speaking, he argued, the Muslims proved better forest officers.[33]

Two Indians testified before the 1888 committee and took a quite different view of the matter. Mian Moti Singh, a Rajput and a forest ranger, complained that the highest posts were always occupied by Europeans or officers of British extraction. He argued that the hardships of a forest officer's job could not be justified when there existed no prospect for rising. Mian Nizam-ud-din, president of the Zemindars' Reforming Society for the Punjab, added that Indians would make themselves qualified, seek more education, and be more efficient if they had the motivation of promotion. He volunteered his two brothers as an example of natives willing to accept appointments in the higher grade.[34]

The more prominent members of the Forest Department were just as divided. Baden Powell, formerly the Conservator of the Punjab, warned that he had "never seen a native forest officer whom he would entrust with the control of a forest division" though a few men may be found among the "manly races" of northern

---

[31] *Public Service Commission*, 8.   [32] ibid., 9.   [33] ibid., 12.   [34] ibid., 15.

India (presumably, Muslims). Mr. E. S. Carr, Deputy Conservator of Forests, Lahore, had never seen a native who was "fit to hold a controlling appointment." Others before the committee pointed out that no broad-based representation of Indians could occur without further reforms in Indian society, because the Indians with the requisite schooling were overwhelmingly Brahman.

J. Campbell-Walker, who joined the Indian Forest Department in 1865, stated that he had no bias in favor of either European or natives for the upper echelons. As Conservator of Forests for the Madras Presidency, he attempted to get the best men possible and felt that "we should endeavor to obtain the best men we can at the cheapest price, irrespective of nationality or race distinction." In fact if all factors were equal, he said, preference should be given to natives, whether of British, European, or Asian extraction. The department could afford the best men as long as the government did not demand a surplus of revenue immediately or in the near future.[35]

A 1915 report of the Royal Commission on the Public Services in India, issued only four years before the act of 1919, revealed the same mix of concerns that arose in the 1880s, and also a similar general preference for Europeans in top positions. Wilhelm Schlich, Professor of Forestry at Oxford, testified that Britain needed to send out men to India of a certain (upper) class "if Great Britain was to keep India . . . and if she did not do that the greater the danger of her not keeping India for ever." He felt that Latin and Greek were helpful for foresters, and that university training was superior to specialized schools. At the time he spoke before the commission the forestry school at Cooper's Hill had recently transferred to Oxford. While he would not exclude Indians from the upper grades, "he would make sure that they were fully qualified to enter it."[36]

The opinion of Berthold Ribbentrop became, in effect, the position of the Forest Department up until 1919. Ribbentrop perhaps felt especially inclined to an open system, but an open system without quotas or preferences – one based on merit, however the outcome. His inclination may be due to the fact that, though he became a British citizen, he was a Hanoverian by birth and emigrated to the empire for opportunity. Educated as a forester in Germany, he felt an acute sense of dislocation when Prussia annexed his home, the kingdom of Hanover, in 1866. Though he passed an examination for the Superior Forest Service of Prussia, he left soon after for India and took a post on the recommendation of the Inspector General of Forests, Dietrich Brandis. As Conservator of Burma, then the Pegu Circle, he earned a reputation as a competent, hard-headed, and eccentric administrator. Later, after serving as Conservator of the Punjab, the government appointed him as Inspector General of Forests, following Brandis' retirement.[37]

Ribbentrop's opinion carried special weight and set the policy for decades to come. He argued that one "cannot accept, without a direct assurance to that effect, that the Government is bound to provide employment for educated natives because

[35] ibid., 14, 48, 49.   [36] *Appendix to the Report of the Commissioners*, 97, 99.
[37] *Public Service Commission*, 17.

they are educated natives. It is, in my opinion, the duty of the Government to provide the State with the best servants for each kind of work, and to pay them at such market rates as will ensure the maintenance of an efficient service. Every other consideration must land us in uncertainty and doubt." Though this position did not result in the aggressive placement of Indians in positions of influence, it clearly did not bar some from achieving promotion.

Education proved another barrier to Indian promotion. Training for the top positions required a thorough background in the sciences, and this was best attained in Britain and Europe. In addition, the working forest models in Britain and Europe were far more compact, allowing a student to study sawmills, tramways, slides, and forest roads in a single forest, whereas the sheer size of most of the forests in the empire made training outside Europe impractical.

But even education did not erase the opposition to native advancement. Ribbentrop complained that while many young gentlemen of native extraction were "at the school active cricketers and runners," they nonetheless gave up exercise to "grow fat and indisposed to hardship" once in the forests, with a fear of tigers and fever, pining away for life in the city. Opposition to placing Indians in higher positions ran deep before 1919.[38]

What manner of education did European, and some Indians, receive in England and Europe? At Cooper's Hill in England students underwent training for two years and two months, including a stay in Nancy, France, for practical studies. The Royal Indian Engineers' College at Cooper's Hill, a college that trained engineers for service in India, became, with the addition of a forestry degree, the first school of forestry in Britain or the British Empire. Students in good health and with character references were required to pass an examination for admittance that stressed science and mathematics. The studies at Cooper's Hill, and later at Cambridge, Edinburgh, and then Oxford, followed a similar line to this eighteen-month syllabus at Cambridge.[39]

1. Principles of Forestry
2. Forest Botany, Part I
3. Silviculture
4. Forest Entomology, Part I
5. Forest Engineering and Surveying
6. Geology
7. Forest Management
8. Forest Botany
9. Forest Utilization
10. Forest Protection
11. Forest Entomology

The recruits at Cooper's Hill and later at other schools took the competitions held by the Civil Service Commissioners for men aged 17–20. In 1903 the Cooper's Hill committee recommended closing the forestry school and moving it to a major

---

[38] ibid., 18.   [39] ibid., 27, 28.

academic institution like Cambridge or Oxford, so the students could get a well-rounded science education along with the forestry training.[40] It also recommended that students proceed directly to India for further practical training at Dehra Dun, skipping the traditional European stay. The reason for this suggestion, which was ultimately adopted, relates directly to the issue of how the environmental innovations in India – though drawing heavily on European forestry – differ substantially from European forestry and mark the beginning of conservation.

The committee stated that

It has been also asserted that the continuance of the forests of India depends, and must in the future depend, mainly on natural regeneration, and that in High Forest the system of selection is, owing to climatic conditions, necessarily adopted over by far the largest area. The result is, according to Mr. Eardley-Wilmot's views, that during the practical training in Germany the probationer applies himself to the minute study of systems of forest management which may never be put into practice during his service, and that though acquaintance with these systems is desirable this prolonged study is far from being essential to the Indian forester.[41]

Much of the area that British officials set aside for forest management was not in fact full forest, but scarred or denuded land surfaces that suffered from massive overcutting, grazing, soil degradation, fires, and broken canopies. Plantations could not redress the balance for a land area the size of India. Accordingly, foresters protected large areas from abuse while natural regeneration restored the forest.

Brandis had pointed out early on that the reserves when first demarcated "were no better than . . . vast extents of blanks with here and there groups of scrub and trees." Fires in the hot season burned vast areas of jungle, while annual burning by natives ravaged deciduous forests. Because of the vast size of the land areas to be managed, Eardley-Wilmot, the first director of the Dehra Dun school, felt he needed macro, not micro managers. He taught students at Dehra Dun the inadequacies of the European management systems and argued that the British would best accomplish their practical training, not in France, but in Dehra Dun.[42]

The committee also recommended no monopoly be given to any one school. The Indian Forest Department and other colonial forest departments should accept students studying forestry at other schools. Additionally, the program should be broad-based enough to allow students to find gainful employment if they were not selected by the Indian Civil Service or other colonial governments.[43]

At Edinburgh, E. P. Stebbing, Professor of Forestry, described the program as consisting of one year of pure science subjects, and then two more years of applied science – Botany, Forest Mycology, Forest Zoology, Forest Engineering, Surveying, Forest Chemistry, and Geology. This program ensured that each student

---

[40] *Report of a Committee Appointed by the Secretary of State for India to Enquire into the Recruitment and Training of Probationers for the India Forest Service* (London, 1908), v.
[41] ibid., xii.   [42] Brandis and Smythies, *Forest Conference, 1875*, 4.
[43] *Training of Probationers*, xiv, 180.

also learned the manual part of forestry labor, such as working in a nursery, planting, felling, thinning and other hands-on applications. Then, in addition to their theoretical training, students would work in a British forest and then go to France or Germany for a final course.[44]

From 1935 to 1947 forest officers of European extraction had no doubt about the coming independence and worked to integrate Indians in a smooth transition to an all-Indian service. While the Second World War interrupted independence for five or six years, all officers knew it was coming. In fact, as expressed by the 1935 joint committee, the British public had widely acknowledged that "the plea put forward by Indian public men on behalf of India is essentially a plea to be allowed the opportunity of applying principles ... which England herself has taught." Little bitterness remained among British forest officers in India concerning independence. Apprehension about corruption perhaps haunted officers as they sought opportunity in other colonial services or retired to England, but there was also a generous fund of optimism and goodwill.[45]

Nature as inventory: market valuation of forests

The environmental revolution in India involved a new evaluation of the worth of the forest. Nature, in this case nonagricultural land, had been seen primarily as waste and as an obstacle to civilization. Part of the agenda for Brandis, Schlich, and Ribbentrop involved convincing the government of India and the ruling class of the Indian Empire that the value of forests was underestimated.[46]

Railways had exacerbated this underevaluation. With the spread of rail, forests were increasingly made available for marketing, while the demand for both sleeper ties and fuel to fire the steam boilers devastated adjacent forest areas. The railways coupled with increased uniformity of British law increased internal trade and commerce, which in turn raised the number of flocks, herds, and agricultural productivity. With some notable exceptions, British rule decisively ended the Malthusian cycle of demographic expansion and starvation that had characterized Indian society.[47] This increase in wealth and wellbeing brought about an increase in population, consumption and, inevitably, the consumption of wood.[48]

[44] Royal Commission on the Public Services in India, *Appendix to the Report of the Commissioners*, vol. xv (London, 1915), 112, 107, 76, 77.
[45] Session 1933–1934, *Joint Committee*, 26–27.
[46] Dietrich Brandis was the first Inspector General of India and was considered by his successors, Schlich, Ribbentrop, and Stebbing to be the founder of Indian forestry. Dalhousie summoned Brandis from Germany to implement an India-wide forest policy. See "Sir Dietrich Brandis," *Indian Forester* 33 (1907): 284, 285.
[47] "Administrative Report on the Railways in India," 229.
[48] See Commonwealth Institute, Julian Danvers, *Report on the Railways of India*, 1861–1862, para. 49; *Report on the Railways of India*, 1862–1863, para. 34; *Report on the Railways of India*, 1863–1864, para. 42; *Report on the Railways of India*, 1865–1866, paras. 59, 60, and 61; *Report on the Railways of India*, 1866–1867, paras. 50 and 51.

The need for railways pressured the government of India to inaugurate forest reserves. Julian Danvers, India's government director of railways, wrote a series of reports to the Secretary of State for India in which he argued that forest reserves were necessary to supply fuel where coal was scarce. Additionally, he argued that forests were essential in the construction and upkeep of rail lines.

Peasants, merchants, and government officials undervalued the forests, reaping a short-term profit that destroyed the long-term value of mature timber and varied forest products. For instance, earlier in the century, under British rule, large "settlements" had been made in Bengal and the Punjab which transferred to landholders and merchants thousands of square miles of forest lands. These forests were subsequently cut in a single harvest, with no replanting effort, no miscellaneous forest products, and minimal revenue to the government.[49] This practice the new forest department ended.

The interference with the reproduction of trees often proved disastrous to the entire economy of a region.[50] Grazing proved especially injurious, because seedlings would be eaten – usually by goats – before they could attain a safe height. Thus, even when a forest had not been harvested no new growth replaced older trees. The loss of timber and miscellaneous forest product resulted only in the extraction of grass, which could possibly have been attained elsewhere, or in any case did not compensate for the amount of lost firewood and timber to the local village.[51]

To remedy the underevaluation of forests, Brandis argued for the elimination of the "lump sum lease" that allowed the lessees to pay a fixed fee per annum and then exploit the forest area for maximum profit. This system contained no provisions for replanting and no incentives for maintaining a sustained yield for the future. Nor of preserving strands of forest on headlands and banks of streams and rivers, or of preserving undergrowth to protect soil from erosion.[52]

A working plan usually included a royalty on timber, which to the modern ear sounds like exploitation of nature and distinctly unecological. But a royalty system reduced waste and senseless devastation and made it possible to maintain a yearly revenue for the forest department as well as the enforcement of sound environmental policy. Compared to the lump sum system, working plans marked a significant milestone in environmentalism.

Ribbentrop described the devastation of undervaluation in Burma, where

for instance, it was quite usual to fell a tree in order to collect leaves for cigarette wrappers. Areas amounting to thousands of square miles were everywhere annually destroyed by axe and fire for the sake of reaping one or perhaps two, often precarious, crops of cereals sown in the ashes; cattle and even goats grazed unchecked, and forests were fired to provide more extensive grazing grounds. Boundaries of forest property, though frequently shown on the maps and sometimes indicated on the ground, had no practical meaning; and the forests

---

[49] Ribbentrop, *Forestry*, 60.
[50] "Grazing and Commutation in the C.P.," 415; "Grazing in Forest Lands," 235; "Effects of Grazing on Forests," 283.
[51] Ribbentrop, *Forestry*, 161–162.   [52] ibid., 211.

inside the boundaries were all maltreated in the same manner as those outside. These abuses, which were dear to the Indian raiyat and clung to with the tenacity and ultra-conservatism of the peasant, had to be stopped.[53]

The realization of market value also meant the forest itself had to be proclaimed, demarcated, regulated, and policed. The British colonialists were imposing the forest as an idea on a population reluctant to adopt environmental practices. To see the market value of the forest meant the imposition of a "forest conscience" on the minds of the local inhabitants, as well as the merchants and the government of India. Imperialists persuaded the people of India to view a forest as a potential treasure house rather than an obstacle to civilization.

Brandis imported the German practice of joint government and private ventures to attain this market valuation of forests. In Burma, for instance, to encourage the use of ironwood (*Xylia dolabriformis*) for railroad ties the forest department built a steam sawmill near a large stand of ironwood and sold the sleepers on the open market to prove their worth. When the demand for sleepers grew, the forest department leased out the mills, along with contracts on the forest. With this method another wood, the Andaman Padouk tree (*Pterocarpus*), became the focus of private merchants, who followed Brandis' suggestion to sell the wood as sleepers. Thereafter two undervalued woods produced revenue to the department and placed less strain on the exploitation of teak.[54]

Twelve hundred different species of trees were known to botanists to exist in the Indian forests, but in the 1860s the harvest relied upon teak, deodar, and sal.[55] Gamble's *Manual of Indian Timber* gave the world a glimpse of the potential uses that a forest had, with regards to the technical quality of timbers exhibited in the Paris Exhibition of 1878.[56] The first step of multiple use proved to be a market evaluation of the forests and a sense of the variety of uses that they could serve.[57]

[53] ibid., 146. The cutting of seedlings and excessive lopping of new growth Ribbentrop saw as abusive. See "Coppicing of Unprotected Forests," 616.

[54] Ribbentrop, *Forestry*, 206.   [55] ibid., 205.

[56] The Paris Exhibition of 1867, 1878, and the Edinburgh exhibition of 1884 displayed the forest products of European countries, their colonies, and the forest products of participating countries. These exhibitions focused attention on forestry and climate theory, and on the development of forestry innovations in India. They also served as a site of environmental propaganda and transmission, effectively aiding in the formulation of environmental legislation for British colonies outside of India and for other countries who participated in the exhibitions. The United States also hosted exhibitions that promoted scientific forestry. See "The Chicago Exhibition," *Indian Forester* 20 (1894): 31; "At the St. Louis Fair," *Indian Forester* 30 (1904): 433. See also "Edinburgh Forestry Exhibition," *JSA* 32 (1884): 964–965, 1035–1036, 1071–1073; "The Paris Exhibition," *Indian Forester* 25 (1899): 161; "The Paris Exhibition," *Indian Forester* 26 (1900): 366; J. S. Gamble, "Forestry at the Paris Exhibition of 1900," *Indian Forester* 27 (1901): 1. India, along with the other colonies, had individual exhibitions at the Edinburgh exhibition. See *JSA* 34 (1886): 928–929.

[57] J. S. Gamble, *Manual of Indian Timber* (Bombay, 1902). Gamble directed the forest school at Dehra Dun from 1890 to 1898 and published extensively on scientific forestry in a variety of journals, including the *Indian Forester*. See "The Retirement of Mr. J. S. Gamble, MA, FLS, from the Forest Service," *Indian Forester* 25 (1899): 162; Nisbet, "Soil and Situation," 3; "Proportionate Fellings in Selection Areas," *Indian Forester* 28 (1902): 218.

Settlement of claims

After the establishment of the Forest Charter, declaring all waste and nonprivate land state property, the question arose: what is state property and what is not? Various claims rested upon the forests and the failure to expedite these claims in a manner suitable to both the government and the claimant would have doomed the entire enterprise. The settlement of forest lands were conducted on the *raiyatwari* system, in which cultivated land belonged unquestionably to the cultivators. All else not proven to be private or communal came under the mantle of state property, and out of this property valuable forest areas were selected for administration.

Additionally, legal forest acts provided for regulation of forest lands that were privately held, if "such control appears necessary for the public weal, or if the treatment such forests have received from their owners injuriously affects the public welfare or safety."[58] Also shifting cultivation, on either private or public lands, could be banned at the will of the government. These examples of state intervention "proceeded with more or less vigor in the several provinces of the Empire."[59]

When villages or individuals did not hold property absolutely, then the forest department had to define informal rights. Informal rights included such activities as gathering firewood or grazing animals near a village or in a nearby forest. The rights of the permanent villager settlements were deemed "absolutely inalienable," while next to this the right to timber, fuel, and grazing applied variably to the extended state forests. John Locke's principle of intensity of use guided the issue of settlements: the longer the time a person or group used a forest, the more intensively they used it for livelihood, the stronger the claim.[60]

As a rule, village communities claimed any forests within the local area for village use.[61] The British made a distinction between "united" and "ununited villages," where the former were "settled" by the government, which is to say, an area of forest given to the community, and the latter "ununited" villages that did not get a settlement of forest.[62] Though any person could generally attain new land for farming or grazing, the rule inherited by the British government with conquest lent itself to the establishment of reserved and protected forests when the government declared all uncultivated land – nature – as state property.[63]

Provinces fared differently in the allocation of village and forest lands.[64] Thus where the united villages stood close together and the villagers used all the forest lands around them, the entire forest region would be claimed. But along the foothills

---

[58] Ribbentrop, *Forestry*, 114.

[59] "On Forest Settlement and Administration," 24; Ribbentrop, *Forestry*, 120.

[60] *1875, India Wide*, 28, 29.

[61] Tserofski, "Village Forests," *Indian Forester* 18 (1892): 150.

[62] Futa, "On Forest Settlement and Administration," *Indian Forester* 19 (1893): 24.

[63] Saint-Hill Eardley-Wilmot, "Indian State Forestry," *JSA* 58 (1910): 493–502; Gem, "Plea for Protected Forests," 32; Gem, "Protected Forests," 57; Ribbentrop, *Forestry*, 96.

[64] Tserofski, "Village Forests," 150.

of the Himalayas in the Terai and the Duns, where malaria and wild animals abounded, few villages made a claim. There large stretches of the country fell to government control.

In the hills of Kumaun, Garhwal, and Jaunsar no tradition of villagers utilizing communal forest land existed, and so all the forest in the area went automatically to the state, while in the plains of the Punjab both unoccupied forest and desert land came under the umbrella of protection, even though these areas were still used by nomadic tribes of cattle thieves and robbers.[65] When the British annexed upper Burma in 1887 the vast forest areas that had previously belonged to the native prince fell automatically to the government.

The advent of forest laws to fix and settle the rights of the state over nature gave absolute property rights to the government. While Brandis recognized that forest rights had their origin in "old usage and custom," and argued that it was "in no way justified" to set aside the rights of ancient use in the forest, he nonetheless held that the forest laws rightly gave a formula for acquisition and computed all rights to the state, allowing only the settlement of a forest into protected and reserved zones.[66]

Dietrich Brandis headed the second India-wide forestry conference in 1875. Here one forest officer claimed that with the exception of the Sunderbuns, "there is not a forest in any province under the Government of India in which the public at large may cut a stick without sanction." The reasoning behind such control lay in the multiple-use management approach, where "A forest may be said to fulfill its highest function when it produces, in a permanent fashion, the greatest possible quantity of that material which is most useful to the general public, and at the same time yields the best possible return to the proprietor (the state)." With security of title, a permanent demarcation was the next step.[67]

Mapping nature: survey and demarcation

Before any demarcation of a forest area could be accomplished, a detailed map of the proposed area was required. The government of India had utilized the Survey of India to comprehend the general outline of the subcontinent and its resources. The new forest department depended upon this grand survey and materially added to the effort with its own separate survey branch. Foresters relied upon accurate maps not only to give a cartographical frame to the forest areas it placed under protection, but also adequately and fairly to "settle" the rights of claimants, whether government bodies, villages, or individuals.

The first systematic survey of India was undertaken in the 1760s. Later, with the publication of Rennell's cartographical surveys, scientific cartography came

---

[65] Ribbentrop, *Forestry*, 99.
[66] "Gem, Plea for Protected Forests," 123–136; Gem, "Protected Forests," 57; Brandis and Smithies, *Forest Conference, 1875*, 32.
[67] Ribbentrop, *Forestry*, 124.

to replace personal experience and journalistic descriptions of trade and travel routes.[68] Early colonial maps moved from the medieval emphasis on the sensuous and tactile (such as a picturesque drawing of a city) to spatial organization and area representation. For instance, a Mughal or early European map of India might depict buildings or towns that a trader would encounter, or emphasize a town's wealth by the size of its representation on the map. But British mapping efforts, beginning in the eighteenth century, emphasized the accuracy of longitude and latitude – the terrain plotted to correspond with mathematical precision according to a grid.[69]

In 1802 William Lambton began the great trigonometric survey in Madras. George Everest continued his work in 1823 and measured the length of the meridian arc from Travancore in southern India to Dehra Dun in the north.[70] The trigonometric survey gave a statistical inventory of India as inveighed by Lord Wellesley in 1802, when he insisted that a surveyor should by no means "be confined to mere military or geographic information, but his inquiries should be extended to a statistical account of the whole country."[71]

With the great arc completed under Colonel Everest, meridional and longitudinal chains on a grid superseded the system of triangulation. By the time the forest surveys were introduced it became necessary to mark a sufficient number of lines

[68] Matthew Edney, "Mapping and Empire: British Trigonometrical Surveys in India and the European Concept of Systematic Survey, 1799–1843," unpublished Ph.D. thesis, University of Wisconsin, 1990: 1–35. Susan Gole discusses medieval ideas of India in *India Within the Ganges* (New Delhi, 1983) and in *Indian Maps and Plans: from Earliest Times to the Advent of European Surveys* (New Delhi: 1989). T. H. Holdich, a nineteenth-century surveyor of India, described the efforts at mapping the difficult and often dangerous frontiers in *The Indian Borderland, 1880–1900* (London: 1909). A useful selection of salient historical documents relating to the survey of India can be found in R. H. Phillimore, *Historical Records of the Survey of India* (Dehra Dun, 1945); Clements R. Markham's *A Memoir on the Indian Surveys* (London, 1871) provides interesting firsthand knowledge of surveyors, as well as F. V. Raper, "Narrative of a Survey for the Purpose of Discovering the Source of the Ganges," *Asiatic Researches* 11 (New Delhi, 1979), first published circa 1818. Essential for any student of Indian cartography would be James Rennell, *The Journals of Major James Rennell, First Surveyor-General of India, Written for the Information of the Governor of Bengal During his Surveys of the Ganges and Brahmaputra Rivers, 1764–1767*, ed. T. H. D. LaTouche (Calcutta, 1910). See also George Everest, *A Series of Letters Addressed to His Royal Highness the Duke of Sussex* (London: 1839) and H. L. Thuillier and R. Smyth, *A Manual of Surveying for India, Detailing the Mode of Operations on the Trigonometrical, Topographical and Revenue Surveys of India* (Calcutta, n.d.). A modern treatment is Paul Careter's, *The Road to Botany Bay: an Exploration of Landscape and History* (Chicago, 1987).
[69] See Edney, *Mapping*, 52.
[70] Until 1815 there were two Surveyors General, one for Madras and one for the Bengal Presidencies. The trigonometrical, topographical, and revenue surveys, separately administered, were combined into the Survey of India in the 1870s.
[71] For an account of the first measurement of a meridian arc in India see William Lambton, "An Account of the Measurement of an Arc on the Meridian on the Coast of Coromandel, and the Length of a Degree Deduced Therefrom in the Latitude 12[ring]32," *Asiatic Researches* 8 (New Delhi: 1809/1979).

on the grid to enable streams, rivers, houses, railways, canals, roads, fences, and boundaries to be measured and fixed legally. "Circles" were introduced by surveying from a given point with the telescope pointed out from the station, to check the accuracy of the measurements of the property within the meridian of the ring. The plotting of the forest's boundaries were done with field books by surveyors and its accuracy tested on the ground by examination.[72] Foresters further delineated certain trees or footpaths as markers and then presented them on the maps as legal entities. These representations "[were] then handed over to the officer in charge of the leveling to have the levels and contour lines inserted, and finally to the hill sketchers, whose duty ... [was] to make an artistic representation of the features of the ground."[73]

The imposition of clear boundaries aided the concept of the nation-state and its "natural" boundaries. As one British writer declared, "The natural boundary of India is formed by the convergence of the great mountain ranges of the Himalayas," which, if British power is consolidated, "shall have laid down a natural line of frontier, which is distinct, intelligible and likely to be respected."[74] T. H. Holdich, who mapped modern-day Afghanistan, pointed out, "boundaries must be barriers," and thus if "not geographical and natural, then they must be artificial, and as strong as military device can make them."[75]

Brandis used mapping to do more than represent the topography of the forests. He used maps to "facilitate the settlement of questions of encroachment and boundary disputes," and to enable a systematic working of the forests by determining the division of internal lines and compartments. Therefore he wanted maps to contain "no more topographical detail ... than is necessary" to do the job and to save money. Exceptions would be for reserves, which were afforded a full level of protection for the maintenance of fire lines, or where complex forest rights and privileges had been granted to villagers.[76]

Foresters took the business of surveying very seriously. If mistakes were made the surveyor would have to correct the improper readings in his own time, or pay for another to do so. New and young officers were rarely allowed use of the fine instruments, such as the theodolite or the prismatic compass. The work of native foresters was double-checked by a European, with map lines drawn by a native superseded by a blue line showing it had been properly approved.[77]

---

[72] "The Cutting and Upkeep of Boundary Lines," *Indian Forester* 26 (1900): 617.

[73] *Encyclopedia Britannica*, 11th edn, s.v. "Topographical Surveys."

[74] George MacMunn, *The Romance of the Frontiers* (Quetta, 1978), 44. For a description of the natural northern boundary see R. Strachey, "On the Physical Geography of the Provinces of Kumóan and Garhwal in the Himalaya Mountains, and of the Adjoining Parts of Tibet," *JRGS* 21 (1851): 5 7–85.

[75] T. H. Holdich, *Political Frontiers and Boundary Making* (London, 1916), 46. The search for a natural boundary included a "scientific" boundary as well. See W. P. Andrew, *Our Scientific Frontier* (London, 1880); C. E. Biddulph, *Our Western Frontier of India* (London, 1887); H. B. Hanna, "India's Scientific Frontier, Where is It? What is It?," *India's Problems* 2 (1895).

[76] Brandis and Smythies, *Forest Conference, 1875*, 88.

[77] Baden Powell and Macdonell, *Conference of Forest Officers, 1872*, 1–8.

After the establishment of boundaries, the surveyor and forest officer in charge called all neighbors and interested parties to a meeting to announce the new boundaries and invite objections or the statement of claims. These lines, as much as possible, were established along "fixed points," that is already existing roads, paths, rivers, and natural markers. Once all parties worked out the agreements on boundaries, the forester put the agreement on paper for all to sign.[78]

Natural features had to be marked, making the forest maps the most detailed maps available. Lakes, ponds, rivers, canals, marshes, large rocks, mountains, dells, valleys, cattle roads, public roads, railways, and fields all were, usually for the first time in history, recorded. These natural formations themselves served as markers, and when not convenient, foresters set up boundary stones (or erected pillars, poles, or dug ditches) to mark the limits of the state forests. Foresters also recorded the type of forest growth, age of the trees, and condition of the soil. Villages, government buildings, stone quarries, and forest rest houses were also clearly marked.[79]

The value of the timber to be worked was marked on the map. This method of valuation involved specially constructed tools that measured the diameter of the trunk and the height of the tree. In 1872 these instruments consisted of the Kluppe, the Lofor Diameter, and a Looking Glass Hypsometer. The first two measured diameter, with the Lofor Diameter involving two arms that fitted round the base of the tree. The Looking Glass Hypsometer, an optical instrument, measured the height.

A line of men walking through the woods observed every tree – without instruments – and wrote the number of trees down in a journal. Then the instruments measured an average tree and the results were multiplied by the count. If the forest tract proved too large for a count, a section of the forest that the forest officer deemed representational could be used as a sample, and then the result multiplied by the remaining territory. These methods, though simple, produced a fairly accurate estimate of the timber contained in a given area.[80]

The inspector general of forests could order a survey, in consultation with local governments, of those areas proposed to fall under the status of reserved or protected forests.[81] This constituted the start of the forest survey branch of the great Survey of India, with staff members of the Survey of India addressing the mapping of the forest areas "from time to time," as requested by the forest department after 1873, when the forest survey branch was instituted.[82] Though the forest survey branch began modestly between 1873 and 1878, 1,284 square miles were mapped, averaging 260 square miles a year. This increased steadily until between 1893 and 1898, 10,380 square miles had been mapped for the period, averaging 2,070 square

---

[78] ibid., 2.   [79] ibid., 3–4.   [80] ibid., 9–10.

[81] Gem, "Plea for Protected Forests," 123–136; Gem, "Protected Forests," 57; Ribbentrop, *Forestry*, 137.

[82] ibid., 130.

miles per year.[83] By 1899 a total of 41,021 square miles had been surveyed by the forest survey.[84]

Information as power: from survey to working plans

Brandis took over the forest responsibilities in Burma with no working plans in place. He quickly adjudged the lack of working plans to be the ruination of the forest cover, particularly of the teak forests. He accordingly framed working plans that attempted to guard against overexploitation of timber. A working plan's primary utility lay in the estimation of yield and also in the strategy for fire protection and replanting.[85] Brandis, with Schlich and Ribbentrop as assistants, drew up working plans for all the forest areas surveyed. He sought to avoid the "mischievous state of attempting to protect large areas," and to concentrate instead on the "reservation of limited areas free of all rights."[86]

When Schlich followed Brandis as inspector general, an "epoch-making" shift took place in the organization of the forest department, largely concerning working plans. Prior to 1884 the working plans had been drawn up by the inspector general and the officers on the site, who corresponded with the forestry department head in Dehra Dun for suggestions. Schlich, however, laid the foundations for a systematized collection of data after the survey, a process involving input from the local government, the local forest agency, the technical advice of the inspector general and, after completion, a chance for the local government to have input for criticism or recommendations. He gave fair notice of his interest nine years before he became inspector general, at the second India-wide forestry conference chaired by Brandis. His intentions to build on the work of Brandis and terraform the subcontinent have proved so influential that it is worth quoting him at length on his vision of management.

The preparation of detailed working plans for all Indian State forests will require a great many years. It is evident that we cannot manage the property entrusted to us for so long a time without some sort of a plan . . . I may say that all Indian State forests are in an abnormal state, and consequently the main object of working plans is to lead these forests over into the normal state . . . the area of the forests is very large and the staff of officers extremely small. In order to bring all forests within the next few years under the operation of working plans, the latter must be of a simple nature . . . an officer fairly skilled in the work can prepare in one season a preliminary working plan for a forest area of between 100 and 200 square miles in the plains, or an area in the hills reduced in proportion to the difficulties of the country.[87]

[83] ibid., 129.   [84] ibid., 217.
[85] Vagrant, "Located Fellings: a First Step Towards Regular Working Plans," *Indian Forester* 19 (1893): 371.
[86] Ribbentrop, *Forestry*, 135; Brandis and Smithies, *Forest Conference, 1875*, 32.
[87] Brandis and Smythies, *Forest Conference, 1875*, 104.

Schlich then detailed preliminary working plan suggestions that included a general description of the "working circle," its situation, boundaries, topographical description, climate, character of forest growth, population, number of depots, markets, and the potential for subcompartments and blocks. He gave notes on past management techniques, suggested methods of deducting growing stock, for analyzing the condition of a forest, the rate of growth of trees, the calculation of annual yield, and helpful reproductive measures. He also sketched a mode for obtaining minor forest produce. Lastly, he outlined his suggestions for the collection of statistical data.[88]

The implementation of the working plans branch of the forest department marked the first time that newly reserved tracts of land of a vast area had been set aside and effectively operated for timber and environmental purposes. Much of the early work of empire foresters involved the massive task of taking an inventory and creating a workable plan for management. By 1899 approximately half of the area surveyed in India had working plans in place, approximately 20,000 square miles.[89]

Fire fighting

Having won the battle to annex and manage the forests of the Indian subcontinent, fire fighting proved the first order of business. Forest fires presented a terrifying spectacle that destroyed life and extensive property, figuring worse than earthquakes in the imagination of both the public and of Indian foresters.[90] The climate did not help matters, since the hot dry seasons accumulated leaves, grass, and dead branches and a single spark could result in the destruction of thousands of acres.[91]

Foresters debated the desirability of fire control. Some officers argued that certain species required fire to germinate and that the prevention of forest fires would kill those species. Others said that fire burned off forest litter and prevented worse damage when grass burned annually. But this argument, to be resurrected decades later, never gained universal approval in the nineteenth century. Foresters tried to prohibit the annual burn of underbrush by villagers that destroyed, they argued, the ability of the topsoil to retain water and to drain properly.[92]

Forest fires were deemed destructive to the whole household of nature and to the diversity of forest life. At the 1875 forest conference one officer spoke on fire conservancy. He laid out the interconnection between decaying leaves, grass, vegetable matter, and the topsoil. When a fire destroys a forest, it destroys the standing grass, the leaves, and the decaying vegetable matter. Without this cover, he argued, the exposed soil bakes and hardens in the sun and little moisture can be absorbed. This means little food percolates down with the water to the roots of

---

[88] ibid.   [89] Ribbentrop, *Forestry*, 137–138.   [90] "Forest Fires," *Indian Forester* 19 (1893): 422.
[91] C. P. Tuscan, "The Mythology of Forest Fires," *Indian Forester* 18 (1892): 263.
[92] Brandis and Smythies, *Forest Conference, 1875*, 4.

10 Reserved forest of deodar in the Punjab India. The forest rest houses were built by the Forest Department for rangers who often stayed at the same location for many years.

the trees, and the most valuable soil washes into the streams and is lost to the tree as nutrients. Grass fires, he concluded, even when they do not destroy the whole forest, destroy the relationship between dead, dying, and growing plants.[93]

He also argued that the return of nutrients to the soil through a coating of "light porous material" kept the surface temperature even, allowed "falling seeds to germinate," and supplied food to the "innumerable roots of the side roots of the larger trees." This porous coating of topsoil retained moisture and gave it off slowly, with rain percolating into the soil and carrying with it "the food of plants contained in the vegetable coating, laying up a stock of moisture and nourishment which is available when the trees require it." Thus the trees grow up in a "natural manner" and are healthy and sound.[94]

Analyzing the origin of forest fires was something that occupied Brandis from the start. Commonly it had been thought that forest fires occurred when wind blew

[93] ibid., 13.   [94] ibid.

11 A reserved forest in the foothills of the Himalayas in northwest India. These trees, *Betula utilis*, slowed the run off of rain and provided for steady stream flow in the valleys below.

bamboos together, causing friction. When observation proved this idea wrong, many speculated lightning caused most fires in India, striking dead and dried trees. But in only one instance, after years of observation, could Brandis trace a fire to lightning. The most common cause, he concluded, came from burning grass off the forest floor to obtain a crop of young growth for cattle or sheep. Burning the forest floor helped in the gathering of certain fruits, flowers, and nuts. Fire cleared the ground for convenient walking between villages, making the passage safe. Clearing the forest of underbrush protected fields from molestation of forest animals as well. Burning not only started most of the forest fires in India, he concluded, but presented a steep human and political obstacle to foresters, since villagers and jungle folk found forest fires to be clearly in their own interests.[95]

The implementation of the working plans included a prohibition of shifting cultivation, which in turn reduced the amount of forest fires. Brandis, Schlich, and Ribbentrop found areas denuded to a lunar landscape by the practice, with soil erosion so bad that replanting had become impossible. Other areas denuded by shifting cultivation responded to reseeding. Often broadcast sowing – tossing

[95] ibid., 14.

12 Cart road used as a fire line through a sal forest in Oudh, India. Photo by R. S. Troup, 1910.

handfuls of seed evenly over a given area – was sufficient to replant a forest.[96] In the early 1870s shifting cultivation had largely ceased and "extensive growth of young forests sprang up in its place."[97]

The successful eradication of shifting cultivation, which proved so injurious to forests, could not have been accomplished without the new powers of policing that the Indian Forest Service received from the government. The enforcement of the working plans meant the gradual education of the population towards an understanding that forest product would be licensed. This meant criminal prosecution for those who violated the law. Ribbentrop noted that forest department enforcement of the new laws fell between two extremes; the "over-zealous forest officer" who wished to "protect it with strictness," and the "maj-bap," that is, the district officer, who fancied himself the father and protector of his community and became "captured" by local and native interest.[98]

The introduction of a uniform rule of law also made shifting cultivation obsolete, because of the clear demarcation of property rights and the ability to address the courts with grievances. Thus when fires from shifting cultivation spread beyond an intended area and damaged the homes and crops of others, the courts would have to

---

[96] Nisbet, "Soil and Situation in Relation to Forest Growth," 3.   [97] Ribbentrop, *Forestry*, 171.

[98] *Maj-bap* is a term of respect meaning "You are both my mother and father."

13 A 100-foot wide fire line protecting a reserved forest of pine in the United Provinces, India. Photo by R. S. Troup, 1909.

intervene and settle the dispute. This only added weight to the education efforts of the Indian Forest Department in the battle to dissuade the use of shifting cultivation. It also added to the argument made by Ribbentrop, that shifting cultivation was "incompatible with increasing civilization."[99]

While accidents took a considerable toll, the worst abuse of the forests came from the agricultural practice of shifting cultivation – the burning of grass and forest land to clear away vegetation for a new crop of grass.[100] Ribbentrop, an admirer of Indian culture and people, nonetheless acknowledged that "A great majority of our deciduous forests in almost all parts of India came to us in a ruined state . . . the continuous firing and hacking had left a crop of ill-shapen, gnarled and often unsound trees."[101]

---

[99] Ribbentrop, *Forestry*, 149.
[100] Called *kunri, jhum, khil*, and *taungya* depending on the province. This method usually produced only one or two crops of cereals and had been widely practiced throughout the subcontinent. See ibid., 148.
[101] ibid., 171–172.

14 Teak logs in upper Burma, placed in a dry stream, waiting for the rainy season to transport to the depot. Photo by R. S. Troup, 1899.

The fire line, a unique invention of Indian foresters, proved essential to the protection of the forests.[102] A fire line is a line or road cleared of grass and maintained around a forest. For deciduous trees in particular, which burn fiercely and easily, the fire line had to be sufficiently broad to keep a flaming canopy from spreading from one side to another. The width, the method of clearing, the number of forest guards to watch over and maintain the lines, varied from forest to forest.[103]

The width of the fire line also depended on the direction of the prevailing winds – fire lines at right angles to the prevailing wind required more width. A belt of evergreen trees running along the edge aided the effectiveness of the fire line. Fireguards, who watched the fire line, were often paid a low monthly salary with a heftier lump sum at the end of a successful fire season to encourage honesty and hard work.[104]

In the smaller forests of Europe, where the forest cover was limited or clearly broken up by fields, the fire line had little use. But the British administrators could not have offered fire protection over such vast areas of land if they had not used it. Broad swaths of forest were cleared in a linear line through sections of the demarcated forest. The width of fire lines varied according to circumstances.

[102] See O. C., "Fire Conservancy," *Indian Forester* 25 (1899): 56–58.
[103] Brandis and Smythies, *Forest Conference, 1875*, 5.   [104] ibid., 7.

15 A dry slide that transported deodar beams in Kashmir. 1911.

Generally speaking, the more valuable the forest the wider the lines. Streams, roads, canals, railroad track and other clearances sufficed as fire lines as well, providing the forester with a line of defense.[105] The maintenance of such lines relied upon trained foresters, who would lead a team of local forest guards to keep the lines clear of debris. The work consisted of "the cutting of grass, herbs and bushes over miles upon miles of fire-lines" with the lines "constantly inspected" to see that it was done correctly. The grass that grew over the cleared area then had to be burned under strict supervision to see that it did not ignite the forest that the line had been constructed to protect.[106]

The dry season was most onerous to the forester.[107] As Ribbentrop colorfully described:

what that means under the blazing sun of April and May, followed by a stifling night, can only be imagined by people who have lived in the tropic plains of India and can be realized only by actual experience. It is needless to say with what sense of delight the forest officer out here greets the first showers of the monsoon and sees his fire-traces covered with green sprouts of new vegetation.[108]

[105] "Open Fire Lines in the Coorg," *Indian Forester* 28 (1902): 111.
[106] Ribbentrop, *Forestry*, 55.
[107] L. E. A., "Open Fire Lines in Coorg," *Indian Forester* 28 (1902): 110.
[108] Ribbentrop, *Forestry*, 156.

The Bori Forest in the Central Provinces was chosen in the 1860s by Brandis to constitute a representative forest, first restored and then protected by fire lines.[109] Following the success of this forest, administered at the time by Major G. F. Pearson, the North-Western Provinces introduced the same methods, then Bombay, and gradually the rest of India.[110] By 1881, 7,000 square miles were protected, and by 1899 the area protected by fire grew to 32,000 square miles, a feat nowhere else duplicated in the nineteenth century.[111]

Was this protection of forest worth it? Foresters argued for more than the protection of timber, soil, and climate. They argued from humanitarian grounds. One forester, Captain Wood, wrote that the "poor man in cities has to spend about one-eighth of his pay to buy wood to cook his food," and that railway sleepers from Europe, iron girders and beam from England, all costly and hard to transport, were used for public buildings instead of Indian timber beams. This cost to the poor man and to society at large was because the "destruction of wholesale firing of forests" had not been stopped. It was also caused from "the natural desire of rulers not to interfere with the customs of the people." This timidity, Wood argued, must end. Why? Because of the "future interest of the population in the vicinity of the forests" and of the "community in general."[112]

Grazing

The reservation status of a forest in British India meant an imperfect forest to professional foresters. This was because the settlement of rights for a reserved forest allowed for grazing and logging, as well as extraction of other forest products.[113] Though many "a forest officer's heart may bleed that he cannot bring all his forest to that state of perfection" he desired, the reservation status proved to be the most practical model of forestry, and one that other countries adopted.

Grazing is dangerous to a forest because cattle and sheep eat the seedlings of trees, damaging the reproductive potential. Nevertheless, Brandis, Schlich,

---

[109] C. P. Tuscan, "Forest Administration in the Central Provinces," *Indian Forester* 19 (1893): 45, 312; C. P. Tuscan, "Forest Administration in the Central Provinces," *Indian Forester* 17 (1892): 373.

[110] "Pioneers of Indian Forestry: Colonel G. F. Pearson," *Indian Forester* 30 (1904): 293–298.

[111] A narrow stream of protest against fire protection policies did emerge in the Forest Department however, and would grow to dominate fire discourse later in the twentieth century. See H. S., "Too Much Fire-Protection in Burma," *Indian Forester* 22 (1896): 257; H. C. Walker, "Fire Protection in Teak Forests in Lower Burma," *Indian Forester* 28 (1902): 293–298. A spirited answer to this protest can be found in F. Gleadow, "Forest Fires," *Indian Forester* 29 (1903): 93–98; A. W. Lushington, "The Necessity for Fire Protection," *Indian Forester* 30 (1904): 472. By 1905 a marked moderation of tone can be detected that attempted to balance both the prevailing and minority view toward forest fire and its effect. See "A Summary of Observed Results of Fire Protection in Reserved Forests," *Indian Forester* 31 (1905): 421–423.

[112] Brandis and Smythies, *Forest Conference, 1875*, 14.

[113] "On Forest Settlement and Administration," 24; "Grazing and Commutation in the C.P.," 226; "Grazing in Forest Lands," 235; "The Effects of Grazing on Forests," 283.

Ribbentrop, and Stebbing all approved its use. Why? Because grass formed the basis of much of the Indian peasant economy. The Indian government had to allow grazing rights; they constituted a local privilege that made palatable the reservation of large parts of India into protected forests. The forest department had originally attempted reserved forest without grazing rights, and the result proved politically ruinous.[114]

Though the nude hills of the Deccan convinced foresters of the profound danger of grazing, administrators incorporated grazing into the settlement of the reserved forests.[115] This resulted in a less pristine but politically palatable forest system. Valuable income also accrued from grazing, since the forest department licensed the grazing of animals to collect fees. The revenue and political rewards allowed for the protection of vast tracts of land that would have otherwise been lost if a more pristine forest protection had been attempted.[116]

Logging, roads, and soil protection

Methods of logging affected the environment profoundly by determining the efficiency of timber extraction, the erosion of soil, and water quality. Efficient logging meant that each tree species would be treated according to its own peculiar characteristics and replanted according to the requirements of its own reproductive needs.[117] Sandal, juniper, deodar, blue pine, pine, teak, evergreen (fir and cedar, among others) each required special methods of extraction and replanting. Once cut, timber had to be dragged over the surface of the forest, often to a water slide and then a stream. The seasonal torrents from the hills provided efficient streams for carrying out pine and other smaller logs.

Ribbentrop planned a number of slides that protected the soil from erosion. First constructed by Mr. McDonnell, Conservator of Forests in Kashmir, slides were able to float logs up to 365 cubic feet for over 3 miles. Sleeper slides for the railway were in place in the Jaunsar Forest in the North-Western Provinces. However, not all forests were accessible by streams, and the development of the railroad provided the means for further exploitation of timber. While at first this accentuated deforestation, the railway did provide the means of a revenue-producing forest that could be managed in the new scientific forestry manner.[118]

Typically, timber in India had been cut by an ax, dragged by elephants to the hill streams, pushed down the stream until the stream grew large enough for the timber to be formed into rafts, and floated to a large town. There purchasers then processed and cut the wood for the domestic or export market. This resulted in a large loss of wood – often with 85 percent of the original tree stem wasted and

---

114 Ribbentrop, *Forestry*, 161.
115 Nisbet, "Soil and Situation in Relation to Forest Growth," 3.
116 Ribbentrop, *Forestry*, 144, 145.
117 "Proportionate Fellings in Selection Areas," 218; "Improvement Fellings," *Indian Forester* 33, (1907): 415.
118 "Scientific Forestry," 89.

not delivered to market – the erosion of soil, and the degradation of the water supply.[119] With the erosion of soil, flooding increased, replanting in some cases became difficult, and water supplies for drinking, both by animals and humans, became problematic.

The solution to these problems lay with the introduction of the saw, the water slide, and selective cutting. Unlike an ax, a two-person saw cut the tree closer to the ground and wasted less timber. Water slides, unlike a sled, did not erode the soil. And selective cutting of valuable trees still left the forest largely intact. Poorly cut wood also ruined the wood for many purposes and necessitated more felling. Cleghorn, Conservator of Forests in Madras, first introduced the crosscut saw in India to cut waste. Brandis introduced into the forest itself saw machinery where "the forest . . . contained mature timber in compact masses" and where foresters found laborers scarce. For this reason, the forest department facilitated the development of both efficient logging and transportation.[120]

The effects of deforestation from logging were clearly seen in India. Hoshiarpur, a rich agricultural area in the north, was yearly covered by sand in growing desertification due to logging and then, subsequently, overfarming. As Ribbentrop wrote, "Sufficient proof [exists] . . . that the hills in question were once densely wooded." Other examples in Simla and Madras showed that deforestation led to soil erosion, leaving a rock or sand bed unable to afford either reforestation or agricultural improvement. The spice garden in Kanara near the Ghats had to be abandoned due to the destruction of the forests in the hills.[121]

Soil erosion also affected shipping in canals, rivers, and ports, because the soil that washed off hills so quickly silted up waterways. The hills of the Deccan and Eastern Ghats provided the British with an excellent example of this phenomenon. The first Dutch and French settlers on the Coromandel coast had made regular excursions up the rivers Godavari and Krishna. But by the time Brandis became inspector general these rivers could not support ship traffic.

The first conservator of forests in the Bombay Presidency, Alexander Gibson, observed in 1847 that the waterways on the Malabar coast had silted up within the memory of the settlers, and ships could no longer pull into the harbor. These considerations were on the minds of the inspectors general when working plans for each region were drawn up. The introduction of efficient logging, water slides, and logging roads meant that fewer trees could be harvested, yet more timber could be extracted, while soil, waterways, and harbors were safeguarded.[122] Brandis argued that protecting forests and soil was to protect the whole life system of India: agriculture, forest, soil, streams, gardens, plant life, fauna, leaves – the whole "household of nature."[123]

[119] "Forestry and Water Supply," *Indian Forester* 28 (1902): 106.
[120] Vagrant, "Located Fellings," 371; *Forest Conference, 1875,* 69, 78.
[121] Ribbentrop, *Forestry,* 50, 51, 54, 55.
[122] Vagrant, "Located Fellings," 371; "Proportionate Fellings," 218.
[123] Ribbentrop, *Forestry,* 57; Nisbet, "Soil and Situation in Relation to Forest Growth," 3.

Minor product extraction

Another innovation proved to be the extraction of minor forest product. Instead of shifting cultivation that burned the entire forest area to the ground, or extraction methods that denuded the forest area and destroyed the forest cover by soil erosion and grazing, the new forestry methods provided a sustained yield that made possible the extraction of minor forest product that was otherwise lost. By preserving the "household of nature," more revenue could be extracted from minor forest product, and this proved critical to the forest service budget. Brandis, Schlich, and Ribbentrop all agreed that the way to protect "the forests as a defined organism in the household of nature" lay in holistic exploitation, not the exploitation of select timber species only.[124] Such exploitation played a central role in this new environmental innovation.

Minor forest product extraction can be divided into two categories: local consumption and industrial, with much of the latter intended for export. The products in greatest local demand were bamboo and grass, which the government forests yielded in large amounts.[125] The *Kham Tahsil* system allowed entry into the forest for free, and payment to be made to the forest officer when the material selected had been removed. Rules prescribing a conservative cutting were made, with elephant herds consciously reduced. Though grass had been primarily harvested by grazing, this method allowed only a 30 percent realization of the grass product – the rest was either trodden down or burnt. In many cases cattle grazed free in government forests under the rights of settlement.[126] Other products included rubber, cutch, resin, lac, and fibrous plants that were made into paper.[127] Though Ribbentrop felt the department had been far too lenient with grazing, he wrote in 1899 that "the conditions of the country render any sudden change in the present system impossible."[128]

Empire forestry in the spotlight

Nineteenth- and early twentieth-century environmentalism depended upon a consciousness that saw the state representing the good of the whole. Within this framework in India, a "forest conscience" arose that utilized the market value of trees over and against the needs of individuals, who realized a one-time only, immediate profit. The annexation of "waste" land under Dalhousie – a precedent-setting

---

[124] Ribbentrop, *Forestry*, 1, 62.

[125] "Bamboos: their Study, Culture and Use," *Indian Forester* 33 (1907): 727.

[126] "On Forest Settlement and Administration," 24; O. C., "Punishments for Cattle Trespass," *Indian Forester* 20 (1894): 410.

[127] "India Rubber," *Indian Forester* 22 (1896): 476. An account of forest product revenues up to the First World War can be found in R. S. Pearson, "The Recent Industrial and Economic Development of Indian Forest Products," *JSA* 65 (1917): 487–493; R. M. Williamson, *Reports on the Forests and Lac Industry in Rewah State, Central India*, (Allahabad, 1906).

[128] Ribbentrop, *Forestry*, 215, 216.

act – set the stage for the expansion of state power over almost all the forest areas of India. This step violated the *laissez-faire* claim to unfettered access and initiated the beginning of forest preservation and research.[129]

India, as the following chapters show, stood in the spotlight, with its accomplishments discussed around the world, in Australia, New Zealand, China, South Africa, British East Africa, Canada, and the United States. Its multiple-use, revenue-producing forests allowed public access without destroying ecological viability. This solved both the environmental and political problems that many legislators, botanists, and amateur forestry enthusiasts sought. Empire forestry in India became empire forestry in the colonies, and eventually a broad-based conservationist program that became known in the twentieth century around the world as *environmentalism*.

[129] "The Aims and Future of Forest Research in India," *Indian Forester* 33 (1907): 507.

5

Empire forestry and the colonies

The export of forestry ideals and methods from India to the other British colonies is surprising because India, though the most prized of the colonies outside the self-governing dominions, did not host the central institutions that provided the expertise for the empire. Britain clearly filled this role. But because most of Britain's own forests were in private and not state hands, because an abundance of coal had replaced timber, and because, paradoxically, Britain relied on its own colonies to supply such raw materials as timber, Britain did not develop its own state-sponsored forestry program until after the Second World War.[1]

The other colonies looked to Indian foresters for inspiration and expertise because India, under Dalhousie's Forest Charter, was the first to develop and expand state forestry programs. Also, certain issues had sharpened the interest of colonial administrators and their publics in Indian forestry innovations. Some of these issues, which have been already discussed, included a growing interest in the link between climate and deforestation. Timber shortages throughout the empire increased speculation that the supply would soon run out.[2]

The factors that led India to adopt a full-scale environmental program, with the reservation of forest areas, also affected other sectors of the British colonies, including South Africa, Australia, New Zealand, and Canada. "To India," R. S. Troup lectured forestry students in the 1930s, "belongs the credit of having been the first part of the empire to adopt a rational policy of forest conservation and development." There is no longer any doubt, he asserted, as to "the wisdom of Lord Dalhousie's forest policy," because the result has justified the policy. "Instead of worthless tracts of waste land, India now possesses in her forests a property of steady increasing value" for the teeming populations that rely upon it.[3] Facing the same twin problems of timber shortage and climate concern, "the Colonies

---

[1] Pearson, "Teaching of Forestry," 422; P. L. Simmonds, "Past, Present, and Future Sources of the Timber Supplies of Great Britain," *JSA* 33 (1885): 102–124.

[2] "Insufficiency of the World's Timber Supply," *Indian Forester* 27 (1901): 397, 435; "Timber Needs of the Empire," *Journal of the Royal Empire Society* 20 (1929): 465–466.

[3] Robert Scott Troup, *Colonial Forest Administration* (Oxford, 1940), 6, 7.

may well be regarded as the disciples of India," with forest legislation modeled on that of India. Officers trained in India visited, advised, and set up the forest administrations throughout the British Empire in the late nineteenth century and early twentieth century, and trained the large proportion of forest officers engaged in empire forestry.[4]

But other issues, some peculiar to a region, provided the spur to imitate Indian forestry methods. In Canada the Dominion government saw the need to plant trees on the vast prairie region to conserve soil and invite settlers. In Australia lush eucalyptus forests on the coast lands were disappearing, and the stark landscape of the interior threatened to become the landscape of the coast land as well. In New Brunswick the hemlock spruce, whose bark local businesses used for tanning, was rapidly disappearing. In New Zealand soil erosion focused interest on forestry, with deforestation raising the specter of vastly reduced agricultural land. In various parts of British Africa shifting cultivation threatened not only forests but water supply, a horrendous portent for a region threatened by severe and deadly droughts.[5] Meanwhile the influential *Journal of the Society of the Arts* and, later, the *Indian Forester*, as well as a few others, kept up a constant drumbeat of reports on the forestry conditions and needs of various parts the world, most in the empire, but many not, such as China, Italy, Brazil, Russia, the United States, Nicaragua, Argentina, Chile, Asia Minor.[6]

---

[4] ibid., 8.

[5] See "Colonial Forestry," *JSA* 24 (1886): 929. This article is an abstract of a paper read by W. Fream at the Colonial and Indian Exhibition, 1886.

[6] "The Forests of Russia," *Indian Forester* 28 (1902): 323; "Destruction of Greek Forests," *Indian Forester* 28 (1902): 446; "Outlook for Forestry in the Philippines," *Indian Forester* 28 (1902): 39; "The Forest of Uganda," *Indian Forester* 28 1902): 421; "West Indian Timbers" *Indian Forester* 28 (1902): 449; "Forestry in America" *Indian Forester* 29 (1903): 323; C. F. Muriel, "Forest Exploration in the Bahr-el-Ghazal (Sudan)," *Indian Forester* 29 (1903): 401; "The American Bureau of Forestry," *Indian Forester* 29 (1903): 258; "Ceylon Forest Report for 1901," *Indian Forester* 29 (1903): 38; "Notes from an American Forest Reserve, *Indian Forester* 29 (1903): 320; "President Roosevelt on Forestry," *Indian Forester* 29 (1903): 301; "Working of State Foresters in Russia," *Indian Forester* 29 (1903): 482; "The Forests of Finland," *JSA* 27 (1879): 787; "Pacific Forests and Rainfall," *JSA* 33 (1885): 1071–1073; "Forest Distribution of the United States," *JSA* 34 (1886): 1091; "The Forests of Tunis," *JSA* 34 (1886): 980; "The Forests of Tasmania," *JSA* 39 (1891): 672; "The Forest Products of Madagascar," *JSA* 39 (1891): 797; "The Forest of Zululand," *JSA* 39 (1891): 833; "Forest and Mineral Wealth of Brazil," *JSA* 39 (1891): 933; "Forest Products of British Guiana," *JSA* 41 (1893): 833; "Forest Trees of Nicaragua," *JSA* 42 (1894): 777; John Gifford, "Forest Fires in New Jersey," *JSA* 44 (1896): 781; "The Mines and Forests of Syria," *JSA* 52 (1904): 228; "The Forests of Asia Minor," *JSA* 56 (1908): 801; "African Timber," *JSA* 63 (1910): 487; "The Forests of the United States of North America," *Indian Forester* 19 (1893): 27; "Progress Report of Forest Conservancy in Ceylon for 1891," *Indian Forester*, 19 (1893): 73; Gifford Pinchot, "The Forest of Ne-he-ha-sa-Ne Park in Northern New York," *Indian Forester* 19 (1893): 280; D. E. Hutchins, "The Forest of Natal," *Indian Forester* 13 (1892): 205; A. S., "Descriptive List of the Timber Trees of the C. P. of Ceylon," *Indian Forester* 18 (1892): 113; "Botanical and Afforestation Department of Hong Kong," *Indian Forester*," 18 (1892): 470; "Forest Destruction and Russian Famine," *Indian Forester* 18 (1892): 366; Pearson, "Teaching of Forestry," 425; "The Ceylon Forest Administration Report for 1893," *Indian*

Even though Britain itself did not serve as the environmental exemplar for the empire, colonial administrators noted that India could provide both the example and the expertise to meet the regional and global concerns arising from deforestation. Colonel G. F. Pearson, a renowned Indian forestry officer, proclaimed before the Society of the Arts in 1882 that "We have now, in India, a fair number of educated foresters who know their work well, and some of whom, at least, are men of high professional attainments . . . It is not too much to hope that the services of some of these men might be utilized" to establish similar programs for the rest of the empire. The scientific forestry expertise these men could provide – though a sophisticated science in itself – could be communicated simply and without mystery. As Pearson put it, "It means simply to observe the action of nature in a forest, and to follow it, or to utilize it to our advantage when we are able to do so."[7]

The International Forestry Exhibition in Edinburgh in 1884 also attracted public attention. Speakers such as J. Michael, a general in the Indian Army and a noted forester, felt an increased "sense of the importance of conservancy . . . growing up in every part of our vast empire." As a concerned public contemplated the forestry problem of the colonies it could "see at once the beneficent results of their having followed India's lead." As early as 1884, India could "Claim the credit of having, by her marvelous successes during the last half-century," stimulated the peak in

Forester 21 (1895): 156; "Report on the Botanical and Afforestation Department, Hong Kong, for 1894," *Indian Forester* 21 (1895): 350; "Deforestation in Russia," *Indian Forester* 21 (1895): 37; "African Bamboo," *Indian Forester* 21 (1895): 39; "The Cedar of Central Africa," *Indian Forester* 21 (1895): 64; "African Mahogany," *Indian Forester* 21 (1895): 120; "The Mongoose in the West Indies," *Indian Forester* 22, (1896): 38; "The Forest and Fauna of British Central Africa," *Indian Forester* 22 (1896): 167; "The Foresters of the United States," *Indian Forester* 22 (1896): 199; John F. Lacey, "The Destruction and Repair of Natural Resources in America," *Indian Forester* 22 (1896): 241; "The Woods of Samoa," *Indian Forester* 22 (1896): 474; C. Sargent, "Forest Reservation in the United States," *Indian Forester* 23 (1897): 147; "The Burma Forest Administration Report, 1896–97," *Indian Forester* 24 (1898): 169; "Forestry in Madagascar," *Indian Forester* 25 (1899): 389; "Madagascar Rubber," *Indian Forester* 26 (1900): 124; "The Harmfulness of Bushfires in the West Indies," *Indian Forester* 27 (1901): 384; "Political Famine (Africa): a Plea for Conservation," *Indian Forester* 27 (1901): 214; "Insufficiency of the World's Timber Supply," *Indian Forester* 27 (1901): 257, 283; "The American Bureau of Foresters in the Philippines," *Indian Forester* 30 (1904): 47; A. M. Burn-Murdoch, "Notes from the Federated Malay States," *Indian Forester* 30 (1904): 458; "Forestry in the Hawaiian Islands," *Indian Forester* 30 (1904): 237; "Re-Afforestation in Italy," *Indian Forester* 30 (1904): 238; "The Forests of Chili," *Indian Forester* 31 (1905): 299; "Forestry at the World's Fair," *Indian Forester* 31 (1905): 112; "Forestry in Indochina," *Indian Forester* 31 (1905): 51; "The Timber Resources of Liberia," *Indian Forester* 31 (1905): 722; "The Devastation of the Forest in West Africa and the Diminution in the Water Supply," *Indian Forester* 31 (1905): 420; Pearson, "In the Uganda Forests," 407; "Proposed Forest Service in the Hawaiian Islands," *Indian Forester* 31 (1905): 412; "Regeneration of the Teak Forests of Java," *Indian Forester* 33 (1907): 284; "Woods and Forests of the Soudan," *Indian Forester* 33 (1907): 261; "Timber in Nigeria," *Indian Forester* 33 (1907): 567; "The Forest of the Ivory Coasts," *Indian Forester* 33 (1907): 747; "Gradual Effect of the Disappearance of the Earth's Forest," *Indian Forester* 33 (1907): 600.
[7] Pearson, "Teaching of Forestry," 425.

"public interest in forest conservancy throughout the colonies."[8] Michael claimed a widespread influence for India on colonial environmental policy as early as 1894.

> Turning to our colonies and dependencies, we see at once the beneficent result of their having followed India's lead. Ceylon has a fully organized Forest Department, consisting of a Conservator, with an assistant, in each province, with a staff of foresters. The straits have a Director of Forests, who is also director of the Botanical Gardens, with various forest officers under him. New South Wales has an establishment of 64 forest officials, at first under a separate director, but now merged with their Department of Agriculture. South Australia has had a successful Forest Department since 1883, consisting of a Conservator and 11 other officers. Victoria has a Conservator, appointed in 1888, with a staff of 28 officers and subordinates. Tasmania has had a Conservator since 1880. The Cape has had a Conservator with four officers under him. Mauritius has had a Forest Department for 15 years, consisting of a Director and 16 subordinates. Cyprus has a Principal Forest officer, with a staff of 44 minor officials.[9]

As the number of countries grew that imported Indian forestry ideas, Indian forestry slowly transformed into "empire forestry." In 1890 Wilhelm Schlich wrote that "Under these circumstances it seems to me essential that the British Empire, as a whole, should endeavor to safeguard against a calamity" of timber shortages. In his *Manual of Forestry* published in 1906 and widely read by forest officers (and even more widely reviewed), Schlich gave some of the basic reasons why the other areas of the empire ought to adopt forestry policies as laid down in India: (1) forests supply wood for fuel and building materials; (2) forests absorb investments and return a profit; (3) forests provide rural employment and income to the working class; (3) forests moderate temperature; (4) forests increase rainfall; (5) forests regulate water supply, particularly aiding in the flow of water during the dry seasons; (6) forests reduce erosion, flooding, and the silting up of harbors; (7) forests produce oxygen and ozone for health and life; (8) forests protect the animal kingdom; (9) forests increase the beauty of a country and satisfy the human thirst for aesthetic surroundings. These basic arguments were repeated around the globe and reprinted in a variety of journals and magazines.[10]

But Schlich called not only for increased forestry policy and practice, but also for an empire-wide integration of the forestry practices pioneered in India. In response the International Congress of Sylviculture was formed, a variety of forestry books were published, particularly by Clarendon Press, and forestry issues took on an increasingly global perspective. Topics once discussed solely in regional or local terms were discussed on a wider basis. It was argued that famine, for instance, could be combated in North Africa, Central Africa, and Asia with forestry.[11] Soil and water flow could be protected over the whole of the earth's surface if forestry

---

[8] J. Michael, "Forestry," *JSA* 43 (1894–1895): 102.   [9] ibid., 103.
[10] W. Schlich, *Manual of Forestry: Forest Policy in the British Empire*, vol. I (London, 1906), 47.
[11] "Political Famine," 214–215.

would be instituted.[12] Though in 1900 foresters were not yet using the term *empire forestry*, the idea of an empire-wide and even global approach to environmental problems was standard.

The First World War completed the final extension of Indian forestry practice to the rest of the empire. This was due to the massive drain on forest resources that the war entailed, the shortages that resulted, and the concerns of national security that arose over timber supply.[13] By the time of the first British Empire Forestry Conference in 1920, foresters spoke of empire forestry as an accomplished fact. After 1920 Indian forestry practice was empire forestry.

But when Schlich wrote his rationale for forest reservations, empire forestry was a proposal, not an accomplishment. "It is proposed," he said, "to apply the conclusions arrived at . . . to the British Empire, insofar as this is possible at present." The total area of the empire in 1906 stood at 12,000,000 square miles, and contained a population of over 400 million. The following sketch traces briefly the development of forest policy in the various regions of the empire, focusing on the appointment of Indian officers and on a series of reports that proved influential to the development of forestry legislation.[14]

Empire forestry in Africa

Next to India, the colonies of British Africa had advanced the most environmentally in the nineteenth century, with the Cape as the launch pad of Indian-style reforms.[15]

---

[12] "Cause and Effect of the Gradual Disappearance of Forest on the Earth's Surface," *Indian Forester*, 33 (1907): 600–601.

[13] E. P. Stebbing, "Forestry and the War," *JSA* 44 (1916): 350–360; Major-General Lovat, "Forestry," *JSA* 49 (1921): 99–113.

[14] There is a surprising paucity of contemporary work on the history of forestry and environmentalism in this period. The best work on Australian forestry is L. T. Carron, *A History of Forestry in Australia* (Rushcutters Bay, New South Wales, 1985); the author provides a brief sketch of national forestry in the nineteenth and early twentieth century. Stephen Pyne's *Burning Bush* (New York, 1991) provides a specialized history of fire in Australia, but little historical framework or broad environmental history. N. B. Lewis also provided a few pages of forest history for South Australia in *A Hundred Years of State Forestry: South Australia, 1875–1975* (Australia, 1975). For South Africa see Richard Grove, "Colonial Conservation, Ecological Hegemony and Popular Resistance: Towards a Global Synthesis," in *Imperialism and the Natural World*, ed. John M. MacKenzie (Manchester, 1990).

[15] There is no broad-based treatment of forestry in the nineteenth century in South Africa, but the *Indian Forester*, and a small assortment of other journals do trace – in fragments – an incomplete sketch of major developments, usually without a larger framework. See James Fox Wilson, "Water Supply in the Basin of the River Orange, or Gariep, South Africa," *Journal of the Royal Geographical Society* 35 (1865); Swellendam, "Conservancy of Forest," *Cape Monthly Magazine* 26 (1878); Hutchins, "Forest of Natal," 205; "The Cape Forest Reports for 1892," *Indian Forester* 20 (1894): 26; D. E. Hutchins, "The Cluster-Pine in South Africa" *Indian Forester* 24 (1894): 125; F. G., "Forestry at the Cape," *Indian Forester* 25 (1899): 384; "The Cape of Good Hope Forester Report for 1898," *Indian Forester* 26 (1900): 589; "National Forestry," *Indian Forester* 26 (1900): 34; D. E. Hutchins, "An Invitation to Indian Foresters in South Africa," *Indian Forester* 26 (1900): 170; "Forestry at the Cape of Good Hope During 1899," *Indian Forester* 28 (1902): 182; "Cape of Good Hope, Report on Forest Administration for the Year 1904," *Indian Forester* 30 (1904): 425; "Afforestation, South Africa," *Indian Forester* 34 (1908): 692.

16 Called "native gardens" by foresters, this former forest in Northern Rhodesia is an example of shifting cultivation. Large trees, too big to fell, were burned at the base and left standing as dead wood. The other forest material was burnt and spread as ashes over the ground for fertilizer.

Though in 1819 a superintendent of lands and woods administered forests at Cape Town and in 1876 the government instituted a forests and plantations department, not until 1881 did the government establish an effective forest department on scientific forestry lines, drawn from Indian expertise. A French forestry officer, Count de Vaselot, superintended the department, with J. Storr Lister and D. E. Hutchins, both from India, to assist, along with various other officers drawn from the Cooper's Hill Forestry School.[16]

Lister introduced plantations that supplemented the sparse natural forests of the Cape. Putting his experience in India to good use, he also worked to reclaim drift sands at Belville near Cape Town. With Lister and Hutchins pointing to the Dalhousie Forest Charter, the dominion introduced Cape Forest Act no. 28, which reflected Indian procedures, particularly the right of the state to demarcate forests as inalienable and to organize forestry officials with police and regulatory powers. In 1906 Hutchins became principal of a newly founded forest school at Tokai (later moved to Saaveld). Then, with the union of 1910, Lister became chief conservator of forests for the entire union, a post he held until 1913.[17]

That South Africa should model itself on Indian forestry practice was early recognized. In 1865 James Fox Wilson wrote an article in the *Journal of the Royal*

---

[16] Schlich, *Manual*, 137.

[17] H. A. Lückhoff, "The Story of Forestry and its People," in W. F. E. Immelman, C. L. Wicht, and D. P. Ackerman, *Our Green Heritage: a Book About Indigenous and Exotic Trees in South Africa, About Trees and Timber in our Cultural History and About our Extensive Silvicultural, Forestry and Timber Industries* (Cape Town, 1973), 25–28.

17 Crown land in Northern Rhodesia during a harvest for firewood. The sheer scale of the harvest indicates the severity of population pressure on forest land. It also illustrates how community forestry and the local harvesting of firewood can be a large-scale operation with severe environmental strains. 1956.

Geographical Society that drew attention to the diminishing water supply in the River Orange basin. In this analysis India figured large, and the arguments used by Indian foresters regarding deforestation and water flow figured even larger. Wilson wrote of "A very noticeable physical fact" that in the 1850s and 1860s "large tracts of country" had been drying up. Springs once fertile no longer flowed, and "desert sucking-places" and "well-filled pools" so vividly described by the missionary Livingstone had completely dried up.[18] Wilson argued that this great change had

---

[18] Wilson, "Water Supply in the Basin of the River Orange," 106.

occurred since the arrival of the Bantu and Europeans in the Cape. He referred often to Robert Moffat, the Scottish Congregationalist missionary, and to Moffat's son-in-law, David Livingstone. Using these accounts, Wilson estimated that the Orange River had once possessed a water flow far in excess of what contemporaries observed in 1865. Much as Indian foresters had compared current conditions to ancient accounts, Wilson used Moffat as well as Livingstone to gauge conditions before widespread deforestation.[19]

The descriptions of drought by early missionaries, who advanced deforestation theories of their own, considerably affected Wilson. He advised settlers to take preservationist action. "Is there any cause," Wilson asked, "besides the interior position of the country and the natural aridity of the soil, which occasions the advance of drought? WE ASSERT THAT THERE IS, and that the effects of that originating cause are controllable, and indeed to a large extent preventable." The answer lay in the destruction of the forest. "THE NATIVES HAVE FOR AGES BEEN ACCUSTOMED TO BURN THE PLAINS AND TO DESTROY THE TIMBER AND ANCIENT FORESTS."[20] To stop this practice would be to stop the drought and increase the rainfall.

India gave the key to understanding the cause of the drought and to finding the solution. Wilson wrote that "Many provinces of India – more especially the Punjab and the Dekkan . . . [may] be adduced in support of the assertions that have been made, the vicinity of hills having become deserted in consequence of the failure of springs following the destruction of woods: but where the digging of canals has been accompanied by the planting of trees along their banks, the departed barrenness has been again transformed into fertility."[21] Likewise the solution for Africa, he concluded, would be to adopt measures as done in other colonies to restore vegetable growth.[22] That India attracted the attention of observers like Wilson as soon as the early 1860s illustrates the remarkable nature of the Indian innovations and gives a preview to the attention it would later receive.

The call for a forest department along the lines of the Indian model grew steadily louder in South Africa. The *Cape Monthly Magazine* printed in 1878 an impassioned plea for a forestry department that would accomplish for South Africa what the forestry department had so "magnificently" accomplished for India. With few resources, the author (signed Forester) argued, "the Indian Forest Department consisted of a small, badly paid, and un-influential set of men." Though the public could see no use for them, the zealous application of sound forestry principles made this department into "a mighty and efficient means of providing the whole country" with an abundant and "never failing" supply of timber for fuel and wood products, for both the domestic and foreign market. "Had the Forest Department not been in existence, India would by this time have been as treeless as South Africa."[23]

[19] ibid., 112.   [20] ibid.   [21] ibid., 122.   [22] ibid., 125, 128.
[23] Swellendam, "Conservancy of Forests," 163.

The author proceeded to illustrate the adaptability of Indian methods for South Africa, describing the work and structure of the Indian Forest Department. In an example of how an Indian-style forest department would benefit South Africa, the author described a scenario: "Let us suppose that 250,000 Deodar sleepers are required for the Railway works, and that these have to be supplied within one year." In India, so efficient was the forest department that the inspector general could write to his subordinates, in this case, the conservators of the North-Western Provinces and the Punjab. He would state the price the railway department was willing to pay for the sleepers, and then the order passed to the deputies in charge of forest divisions, who sent out their assistant conservators to different forests; roads were laid for logging, working plans consulted, workmen sent to log and to replant, lumber transported to mills, and sleepers delivered to the department – all within the framework of a sound, revenue-producing forest department. All this and the department would also save for future generations a forest that would protect climate, soil, and water flow.[24]

Why not take the "admirable plan" of the Indian Forest Department and transplant it directly to South Africa? The "Forester" ended with a call for "establishing a similar system in this country . . . Let the Government employ an experienced man from India or elsewhere, as Commissioner of Forest Conservancy [and] Inspector General of forests."[25] As the employment of Lister and Hutchins illustrates, Indian foresters themselves organized the department and served as chief conservators well into the twentieth century, until empire forestry became uniform throughout the empire.

Lister and Hutchins were the answer to the call for Indian forestry in South Africa. A forest act of 1888 modeled itself on the Madras Forest Act of 1882, and "in this way, the forests which still existed were converted into reserved State forests" with an area of 478,867 acres, approximately 3 percent of the total area of the colony.[26]

The committee that drew up the Forest Act of 1888 called on Hutchins as a witness. It agreed that the present bill being considered was "principally founded" on the Madras Act of 1882.[27] Since the 1888 bill in the Cape Colony is the founding legislation for the modern forest department of the Cape and later, of South Africa, it is worth quoting at some length an interesting exchange between the committee and Hutchins.

Why was it necessary to introduce this stringent legislation [in India]?
[Hutchins] The Forests were gradually being destroyed; the magistrates complained that they were powerless, and could not enforce the preservation of the forests with the laws then existing, and so the destruction continued.
Did that destruction arise from burning or the cutting of wood by people who had no right to cut it?

---

[24] ibid., 164, 165.   [25] ibid., 171.   [26] Schlich, *Manual*, 137, 138, 140–142.
[27] *Report of the Select Committee on Forests Bill* (Cape Town, 1888), 47.

[Hutchins] It arose mostly from cutting. There were fires in certain places, but generally fires in India are not so destructive as they are here in the Colony.

Do your remarks apply to the natural forests?

[Hutchins] Yes. By the time that they had artificial forests in India they already had legislation to protect them.

Are the forests in India difficult to manage in consequence of the excessive population?

[Hutchins] The populations vary very much.

Is a large staff of forest officers kept there?

[Hutchins] Yes.

Can you give the Committee an idea of the strength of that staff?

[Hutchins] Under the Government of India, exclusive of the presidencies of Madras and Bombay, the number of forest officers of superior rank amount to a hundred.

And I suppose the staff of forest rangers is proportionately large?

[Hutchins] Yes.

Are they supposed to patrol the forests?

[Hutchins] Yes.

Can you say whether, under legislation in India, forces of destruction have been checked there?

[Hutchins] Yes. The legislation has finally succeeded.

Finally?

[Hutchins] At first the Acts passed were not sufficiently stringent, and they failed. Successive legislation followed, and finally the Madras Act of 1882, which was found to be effective.

Is there any necessity in this Colony, considering the extent of the forest land, for legislation of that character of the Bill before the Committee?

[Hutchins] I think so.

Will you give your reasons for arriving at that conclusion?

[Hutchins] We have had a constant succession of difficulties during the last five years owing to the want of legislation on the subject. We have failed in our attempts to protect the forests for want of proper laws.[28]

The committee drew up legislation based on the Madras legislation of 1882 and then gave it to Hutchins to personally redraft. While the components were similar to Indian legislation in many ways, included was an interesting innovation by Hutchins – the ability of the government to swap land in its possession with private owners for land that it valued. This innovation is still used heavily by environmentalists today to gain ecologically valuable land.[29]

This was not Hutchins' only innovation. He also advocated that the bill deal with the protection of game. Again the issue is important enough to quote at length.

With regard to the offenses and punishments referred to in Chapters II and IVM, have you any remarks to offer?

[Hutchins] In the matter of demarcated forests an important omission has been made; the hunting and destroying of game has not been dealt with. In the Eastern forests the game has been preserved for the last four years – I believe it is the only place in the Colony where it has been systematically preserved. Without legislation this game will now – is in fact being

[28] ibid., 47, 48, 49.   [29] ibid., 58.

now – destroyed. And what is even more important, native hunts are always followed by
outbreaks of forest fires.

You think hunting and destroying game should be offenses under an additional sub-section?
[Hutchins] Yes. I think that should be embodied in the draft, firstly to preserve the game,
and secondly to preserve the forest; for almost every native hunt is followed by the outbreak
of a destructive fire.[30]

Later Natal forests were added to the list of the reserved, as well as those of
the Orange River Colony and the Transvaal. Hutchins described the indigenous
forests in the northeast of the Transvaal as a "Wood-Bush Range" of evergreens
indigenous to southern Africa and running through the gorges of Table Mountain
to, in intervals, the southern coast of Cape Colony, to Knysna, a mountainous
region with ample rain, to spots in Natal and Zululand, thus cutting across different
political regions and needing the protection afforded by the 1888 act.[31]

A report by the conservator of forests for the colony of Natal in 1891–1892
illustrated the principles imported from India. This conservator made it clear that
profit could not be expected "for some years to come." These principles included
the protection of forests on crown lands, the appointment of native guards under
scientifically trained European foresters, the establishment of plantations, and the
prevention of native burning of woodlands. The report also listed the methods of
surveying using the system of triangulation and the need for clear demarcation,
fire protection, and regulation of timber and minor forest product extraction.

In the British East Africa Protectorate an Indian forester – after a tenure in the
Cape – organized and ran the Forest Department for this less-settled territory. This
Protectorate consisted of a coastal strip with plentiful rainfall and a dry semidesert
interior inhabited by nomadic tribes. The forests grew primarily on the coastal lands
and in the hills surrounding the Rift Valley. Population pressures had removed most
of the forest cover on the coastal strip long before the arrival of the British. The
interests of the colonial office centered on preserving the surviving forests, as had
been accomplished in India.[32]

To this end the colonial office formed a forest department in 1902 and appointed C. F. Elliot to manage the forests. Elliot had retired from the Indian Forest
Department, where he served as conservator of forests for the Punjab, and was
charged, in the words of a Protectorate judge, with making the "native chiefs . . . put
an end to their wasteful practices" because "the prosperity of the country is largely
dependent upon the preservation and extension of the forest land, and I should
make the natives collectively responsible."[33] Elliot began his duties by managing
the forests within a mile on each side of the Uganda Railway, and shortly after

[30] ibid., 60, 61.

[31] F. Schoepflin, *Colony of Natal: Report of the Conservator of Forests*, (Pietermaritzburg, 1892),
5–11; D. E. Hutchins, *Transvaal Forest Report* (Pretoria, 1903), 14, 29. See also Colonial Reports –
Miscellaneous, no. 2, *Report on the Forests of Zululand* (London, 1891); *Colony of Natal: Report
on Forestry in Natal and Zululand* (Pietermaritzburg, 1902).

[32] "Forestry in German East Africa," *Indian Forester* 28 (1902): 372.

[33] D. E. Hutchins, *Report on the Forests of British East Africa* (London, 1909), 79.

18 Hills in Cyprus were denuded by overcutting and goat grazing. Goats ate the saplings and prevented new growth. Photo by R. S. Troup, 1929.

his arrival was appointed Conservator of Forests and given responsibility for the entire British East Africa Protectorate.[34]

Hutchins also served as a conservator. First an Indian and then a Cape forester, he wrote a series of reports to the British parliament on forestry matters and played a major role in hastening the spread of empire forestry throughout the colonies, including British East Africa, where he served as conservator of forests in 1906. His work in South Africa in the Knysna Forest illustrated how the innovations pioneered in India could be adapted to other colonies and, on the basis of his work in the Cape, the British government requested that he visit the forests of British East Africa and then Cyprus, issuing separate reports to Parliament.[35]

After surveying the forests of British East Africa, Hutchins analyzed the applicability of the Indian forest administration to the new territory. Like Wilson, he drew on missionary accounts to identify the destruction of forests. Regarding the practice of shifting cultivation, he quoted Elliot:

In the Wakikuyu country the forests are honey-combed with patches of cultivation; a man goes into a bit of dense forests, cuts down all the trees, and sets fire to them; he cultivates a few crops of sweet potatoes and beans with perhaps some Indian corn, and after a few

[34] Troup, *Colonial Forest Administration*, 337.   [35] Hutchins, *British East Africa*, 26.

19 Pine forest after a forest fire in Cyprus. Grazing from goats prevented regeneration.
R. S. Troup, 1924.

years, when the soil becomes impoverished, he abandons the patch and repeats the process elsewhere.[36]

Elliot's book, *East Africa Protectorate*, written in 1905 and which Hutchins quotes, lambasted nomad tribes for their wasteful clearing of wood. Natives, he complained, used a patch of cleared land for only a short while and then moved away to clear another patch, leaving nothing but scrub brush and grass behind. But if shifting cultivation proved the problem, as in India, then so did India prove the solution. Hutchins wrote that in 1883, when "I came from India and took charge [in the Cape] . . . the forests were in exactly the same position as are the Aberdare forest in British East Africa to-day. The forest was honey combed with native gardens; and the burning and destruction of the forests went on unchecked. To-day all this is changed." After the arrival of Elliot, however, "the forest lies secure within its own boundaries, while the natives are cultivating their gardens outside the forest boundaries in a less barbarous and wasteful manner."[37]

India, Hutchins remarked, "now represents the most complete forest organization amongst English-speaking people."[38] Accordingly, he recommended further improvements that related to the expansion of the department along Indian lines,

---

[36] ibid., 67.   [37] ibid., 67–68.   [38] ibid., 62.

particularly the need to (1) demarcate the forest areas and police them well to keep out infringement, and (2) to exclude the indigenous tribes from the forests by planting a band of white settlers as a buffer zone.[39]

The largest step to be taken, Hutchins argued, was to stop the "alienation" of the forests that exists – that is, the withdrawal from the market of forest lands not claimed as private. Once a forest is alienated, "no means have been devised in the world's history of preserving it. The rights of private property are sacred, the world through" and difficult to reverse.[40] Though a good beginning had been made by Elliot, Hutchins recommended an ordinance that would render forest alienation (privatization) illegal, based upon the Cape Forest Act, no. 20 of 1902.[41]

The East Africa Forestry Regulations 1902, followed by the East African Forestry Ordinance of 1905, set out the formation of the department and modeled itself after similar South African forest legislation, itself modeled on the Madras Forest Act. Organized by Indian foresters, the legal framework copied ultimately from India, the territory joined South Africa as an exemplar of Indian forestry in the colonies.[42]

In southern Nigeria the colonial government organized a forest department with an Indian forest officer, H. N. Thompson, at its head.[43] Then from this post, Governor John Pickersgill Roger invited Thompson to establish a forest department on the Gold Coast along the same lines as the Nigerian department he headed. Accordingly, Thompson issued a report that occasioned the formation of a forest department there in 1909.[44] Since the Gold Coast's economy depended heavily on cocoa plantations, and the reduction of water supply and the increase in dry wind directly reduced the cocoa crop, the securement of catchment forests presented the first challenge to Thompson. The imperial environmentalism initiated in India again provided the model, as the appointment of an Indian forester indicated.[45]

A unique situation not found in India proved to be the ownership of almost all land, including "waste land," not by an indigenous state authority, but by tribes

---

[39] ibid., 68.   [40] ibid., 63.

[41] ibid., 65. This act reads "After the date of the passing for this Act it shall not be lawful, without consent of both Houses of Parliament first had and obtained, to alienate, or grant, or dispose of any servitude upon any Crown Forest which had been formally declared to be a demarcated forest, or any part thereof: and all such alienation, grants, or disposals that may hereafter be made shall be null and void; providing that nothing in this Act shall be taken as affecting existing rights, or as prohibiting the sale of forest produce or the grant of grazing rights in such forests, under and in accordance with regulations made under the Forest Act of 1888, or under the said Act as amended by this Act; and provided that it shall be lawful for the Minister to effect by exchange of lands or otherwise small rectifications of the boundaries of any demarcated forest: and provided further, nothing in this Act contained shall be held to affect the rights of Government to issue permits or licenses, under the provisions of the Precious Stones Act of 1899, the Precious Minerals Act of 1898, and the Mineral Lands Leasing Acts of 1877 and 1883, for prospecting for precious stones, precious minerals, other minerals, or to dispose of land containing such stones and minerals."

[42] Troup, *Colonial Forest Administration*, 337.   [43] Schlich, *Manual*, 144.
[44] Troup, *Colonial Forest Administration*, 323.   [45] ibid., 325.

(the "Stool"). Thus chiefs (who were elected with limited powers and could be "destooled") had to be brought into a power-sharing arrangement. The forest department recruited the local chief as an indigenous conservator who shared authority with the forest department.[46] The Forest Ordinance in the Gold Coast lay down, as Troup argued, "a detailed forest settlement procedure much on the lines of that prescribed in the Indian and Burma Forests Acts," but with the modifications necessary to accommodate a different social and landowning class.[47] By 1933 the forest area of the Gold Coast totaled 13,900 square miles, 2,436 of which were reserved and 4,900 of which were inaccessible.[48]

In Sierra Leone a forest department was formed in 1911. The argument used to justify environmental legislation in India was adopted in Sierra Leone, and provides a clear example of a growing empire-wide justification for the expansion of state power over nature. It also indicates a willingness to compromise with local conditions.

It is the policy of Government to interfere as little as possible with tribal habits and customs and to let change and progress be gradual. The population is an agricultural one, the majority of whom have been accustomed to work in a densely forested country. They have a technique adapted to such an environment. With increase of population due to peaceful conditions there has, in the last twenty or thirty years, been a rapidly accelerating conversion of land from forest to agriculture, and then to more or less barren grassland. This is causing climatic and vegetational changes apt to react violently on the occupations, habits, and custom of the population. In order to mitigate the violence of these, to some extent, inevitable changes, it is the policy of government on the one hand so to change and improve agricultural methods as to lessen the destruction of the forest and on the other hand to conserve existing forests and create new ones so as to preserve or restore, as far as possible, the conditions to which the population is accustomed. It is with the last part of this policy that the Forest Department is concerned.[49]

In Nigeria, widespread deforestation had occurred long before European colonization or industrialization. The Fulani invasion in the earlier part of the century, with the imposition of the Muslim faith, provided stability, rising standards of living, and with it, increased forest exploitation. Vast areas of forest with abundant game, observed by European travelers in the early part of the nineteenth century, largely disappeared in less than 100 years. British imperialism accelerated this process for a short period before the reservation of forest lands on the Indian exemplar.

[46] ibid., 326.   [47] ibid.

[48] ibid., 328, 334. In 1905 the Nigerian colony amalgamated with the colony of Lagos; in 1914 with Northern Nigeria; and in 1938 the Cameroons came under British Mandate, the forests of which all came under the Forestry Ordinance no. 38 of Nigeria. The total area of forests in Nigeria, which supported in the 1930s a population of 20,000,000 was enormous – 243,000 square miles, of which 20,090, almost 10 percent, came under government or native administration Forests Ordinance. See ibid., 328, 329, 333; the 4th British Empire Forestry Conference.

[49] ibid.

But as in the Gold Coast, the forest department kept the traditional system of local control over natural resources. Foresters attempted to educate Nigerians about the need to preserve the forest and, if practicing shifting cultivation, giving enough time between burnings to allow the soil to regenerate. Communal forestry through tribal chiefs enabled foresters to concentrate in the early years on surveying and demarcation. Additionally, the forest department did not project a profit but rather a loss of revenue, as in many Indian forests. This loss of revenue administrators tolerated in the fight against desertification and projected timber and game shortages. This loss continued, from the founding of forestry in northern and southern Nigeria in 1914 (united in 1915) to independence in 1952.[50]

But even though colonies consciously modeled their new forestry departments on the India Forest Department, they often fell short of the mark. This did not go unnoticed by highly placed Indian foresters, who often considered themselves, and were received as, licensed critics-at-large. For instance, E. P. Stebbing, retired from the Indian Forestry Department and as Professor of Forestry and Head of the Department of Forestry at the University of Edinburgh, had much to say on West African forestry practices. Though he agreed that the departments of forestry in West Africa "from the forestry point of view, did not show any abnormal differences from India," he nevertheless made critical points for improvement.[51] Thus "Remarkable as the progress has been during the last decade in West Africa," he pointed out that "the advances made in India in other directions of forest administration, reservation of forests, protection, and so forth . . . had not been reached on the [African] coasts."[52]

One point regarded the definition the forestry departments in West Africa used for "Savannah." Savannah usually was defined in India as a great sal forest with open clearings, the clearings covered by 12-foot tall "tiger grass." Savannah in the United Provinces in India usually did not involve shifting cultivation or the degradation of the ecological balance, but was an indigenous ecosystem of its own. What Stebbing found disturbing in West Africa was the tendency of the forest officials in the forestry departments to designate highly degraded forests as "Savannah" instead of an area in desperate need of reclamation and reservation, thus tolerating, even encouraging, the spread of forest ruination. The so-called savannas were not, he contended, savannas at all, but ruined forests that should be designated as such.

---

[50] These remarks are gleaned from a senior assistant Conservator of Forests for Northern Nigeria, R. W. Fishwick, in *Some Notes on the History of Forestry in Northern Nigeria* (1961). See also H. N. Thompson, Director of Forests, in Colonial Reports – Miscellaneous, no. 51, *Southern Nigeria: Report on the Forest Administration of Southern Nigeria for 1906* (London, 1908), 51–53; H. N. Thompson, *Report Regarding the Irregularities of Rainfall in Nigeria* (Lagos, 1928); Sessional Paper no. 37 of 1937, *Report of the Anglo-French Forestry Commission 1936–37* (Lagos, 1937).

[51] E. P. Stebbing, *The Forests of West Africa and the Sahara: a Study of Modern Conditions* (London and Edinburgh, 1937), 66.

[52] ibid., 233.

Stebbing knew shifting cultivation to be the culprit. In West Africa shifting cultivation consisted of felling an area of forest, burning the trees on the ground along with any brush or debris, scattering the ashes over the area, and then spreading it with grain seed to grow a crop of cereal. This method involved no weeding, little intensive labor, and would produce two or three crops before the farmer moved to the next forest area to cut and burn in the same manner.[53]

Empire forestry had eliminated shifting cultivation over much of its territory and Stebbing intended to see the practice ended in West Africa by removing "Savannah" as the designation of degenerated forests. He wrote in his notes for February 19–25, 1934: "There is no doubt that formerly a vast mixed deciduous forest occupied all the region of this circle ... The more one studies this country the more evident it becomes that the Savannah lands are the degraded condition of a region which was once covered with high forest. If the Savannah was not annually burnt it would probably, or the better parts would, gradually fill up with some of the hardier deciduous forest species."[54]

If these practices did not align with Indian practices then the result would be that "the forest gradually deteriorates over ... a long period of years; the water levels being thereby reduced, thus preventing these degraded forest lands from recovering, unless protection is afforded to them."[55] The "so-called 'Savannah' forest ... was only a varied form of what in India the forest officer termed degrade mixed deciduous forest."[56]

Empire foresters debated the Savannah issue hotly. Whether forest or savanna constitutes the "natural" formation in parts of West Africa is still in question today. While Stebbing argued that it did not, others in the department took issue with him. Some scholars privilege the accounts of local farmers, and claim that African settlers in the Savannah planted much of the existing forest.[57]

Another of Stebbing's concerns involved the southern drift of the great Sahara Desert. From the petrified trees that could be found inside the desert, from the descriptions of the desert's southern outline from writers of antiquity, and also from the observations of village agricultural land that had disappeared into sand dunes, Stebbing foresaw further desertification if shifting cultivation continued. The upper Niger and Lake Chad were particularly threatened. The solution? Grow a forest belt 15 miles deep around the southern perimeters of the Sahara. Within this "belt of high forest ... on a 15-mile depth ... no firing should be allowed, and

---

[53] ibid., 1–2.   [54] ibid., 4.   [55] ibid., 5

[56] ibid., 7. Stebbing took exception with forest botanists who sought an ecological balance that included human firing for the forests. Though "a distinguished forest botanist or ecologist" had lately made the assertion (Dr. T. F. Chipp, Deputy Director of Kew Gardens), he considered the repeated firing of forest areas dangerous "from the practical forester's point of view." The result "is to produce a desert." See ibid., 8, 9.

[57] For a recent polemic against the claim made by Stebbing and most mainstream environmentalists, see James Fairhead and Melissa Leach, *Misreading the African Landscape: Society and Ecology in a Forest-Savanna Mosaic* (Cambridge, 1996).

that cultivation and the pasturage of herds and flocks should be restricted, and eliminated altogether where possible." Otherwise Africa would experience not only increased drought, he argued, but permanent desertification and population displacement similar to the crises that occurred in northwestern India and more recently, in the Midwest of the United States.[58]

Many French forestry officials in Africa agreed with Stebbing's report. A. Aubréville, a French forestry official from West Africa, admitting the difficulty of judging the progress of desiccation due to cycles of wet and dry years, but nonetheless listened to the tales of old men with growing alarm. Wells had dried up, farmlands were abandoned to encroaching sands, rains were less heavy, and the desert moved relentlessly south. This "haunts . . . those who are anxious for the future of Africa," and he referred to Stebbing, "an English Forest Officer" who had alerted "the English press and reawakened the anxiety which has long since been manifested."[59]

After a discussion of Indian forestry methods and organization, Stebbing recommended the use of onetime shifting cultivation, where a farmer could burn a forest, plant a crop, and then turn the area over to forest officials who would reseed with deciduous hardwoods. This method, practiced in India, would thus utilize a practice hard to eliminate by harnessing it for replanting until the practice slowly faded out. Finally, he recommended changes in staffing policy that would make senior conservators of forests more effective in implementing working plans.[60]

New Zealand and Australia

In Australia a series of reports served to prod the Australian government into action and to develop a forestry department along the lines of India's. Though various states of the Australian commonwealth had forestry legislation on the books, these were varied and poorly enforced. In Victoria the 1884 Land Act that provided for the formation of state forests, timber reserves, and proper management, had proved to be a "fine law" without adequate enforcement.[61] Accordingly the governor of Victoria asked in 1887 for a report on the forests of Victoria by "Mr. Vincent, an expert and a trained forest officer of known ability, who served in the Indian Forests Departments since 1873" and who the governor trusted, could help resolve difficulties in the management of the forests.[62]

---

[58] Stebbing, *Forests*, 28–29. See also E. P. Stebbing, "The Threat of the Sahara" *Geographical Journal* 85 (1935): 518.

[59] F. T. Brand, *Proposed Anti-Desiccation Scheme for Northern Nigeria* (Kaduna, 1936); A. Aubréville, "les forêts de la colonie du Niger," *Bulletin du Comité d'Etudes Historiques et Scientifiques de l'Africque Occidentale Française* 19 (1936): 60, 62.

[60] Stebbing, *Forests*, 40–49.

[61] A. Kirkwood, *Papers and Reports upon Forestry, Forest Schools, Forest Administration* . . . (Toronto, 1893), 50–51.

[62] ibid., 50.

20 Fire lookout in Western Australia. 1940.

Vincent painted a gloomy picture of the state forests. He found "useless waste and destruction," with sawmill owners "allowed to go in and cut whatever they chose." Add this reckless management to the "increase in the consumption" of wood, which the natural increase in population indicated, and an extension of mining in forest areas, and the forests would have little future if action was not taken. A bad license system, the absence of professional foresters and of a proper classification of state forest all added, Vincent argued, to the forests "being rapidly ruined."[63]

[63] ibid., 50–51.

21 Firewood cutter tractor in Victoria, Australia. These portable saws opened up many new areas to exploitation.

Like many Indian foresters at large, Vincent made provocative statements meant to catch the public imagination and push the movement forward. He asked if the electors were "prepared to allow the sawmillers and spoilers to devastate the remaining forests, robbing them and their children of their supply of timber and firewood, and risking some of the climatic changes which are traceable to the destruction of forests? Are they prepared to sacrifice a source of large and increasing revenue to the demands of a limited class?"[64] The solution to Vincent was very clear, as the summary of his report which follows shows. "It was suggested that the Victorian Government should secure the services of a fully competent forestry expert, a man like those who introduced systematic forestry into India, who should be directed to go round the colony, see for himself, and then propose what, in his opinion, ought to be done."[65]

But little action was taken after Vincent's report, outside a bewildering array of forestry laws passed separately in the Australian states that lacked the oversight of federal coordination or enforcement. Therefore, at the request of the colonial office, D. E. Hutchins, after successfully aiding the organization of forestry over much of Africa, agreed to tour Australia and New Zealand in 1914 and make recommendations to the forestry departments of the respective states of the Australian

[64] ibid.   [65] ibid.

Commonwealth and New Zealand. As a member of the British Association, Hutchins had – in the words of the chief forest officer in England – "a wider experience in extra-tropical countries than any man living."[66] Hutchins carried a weight of expertise and authority not held by any man of the time, with the exception perhaps of W. Schlich. As a former Indian officer, conservator of forests in Cape colony and conservator and organizer of the forest department in British East Africa, he determined to recommend changes that would make the forestry department in Australia and New Zealand as efficient as it was in India.[67]

The result of his tour was *A Discussion on Australian Forestry*, a monumental tome for forestry literature at this time. His trenchant criticism lectured the forestry department on the essentials of good forestry, from settlement of rights to demarcation of forest areas, demanding accountability and attention to detail. He quoted the conservator of Victoria at the Melbourne Forests Conservators' Conference in 1912 to say "in the short space of 38 years the various governments which have held office . . . have deliberately allowed some 300,000 acres of blue-gum, spotted gum, mountain-ash, messmate and blackwood to be destroyed . . . by axe and fire" and "wasted in the course of what is falsely called successful settlement" of the forests of the country.[68] Hutchins concluded that "over the evil in the other states a discreet official veil is drawn." This evil included conservators with no power, each of whom was "employed mostly like a subordinate clerk in the Lands Department, sorting out paper and taking them up to the minister for his initials." But when the minster of lands "coolly revoked the temporary reservation of another slice of the forest," this same minister boasted of the vast area that had "turned up the development of the country."[69] Thus conservators had little authority to accomplish the permanent protection of forest lands.

Though the mainland states have a horrendous record, Hutchins wrote, yet "Tasmania is the black sheep" in Australia forestry, where most of the accessible forests have been "burnt and destroyed."[70] Protection of the eucalyptus forest "is considerably easier" than in other similar climates where protection had been maintained for years. The solution was clear. "I am in a position to assert positively that control of fires in Australia forests is solely a matter of organization."[71] Forest fires in India were a case in point. "Here the bamboo grows like a huge grass 20 feet or 30 feet high . . . the fire advances in a wall of flame, the bamboo joints exploding the while like pistol shots." But all that is required is that "the whole forest is cut up by broad fire-paths, into squares like a chess board, and when fire breaks out, it is usually possible to confine it to one, or perhaps two or three squares."[72]

---

[66] D. E. Hutchins, *A Discussion of Australian Forestry* . . . (Perth, 1916), 21.
[67] The preface of the Hutchins report, written by an official of Western Australia, mentions that "The government of Western Australia felt that in the interest of the state, which possess so much forest of special character, the opportunity of consulting a forester with such ripe experience and matured judgment was not to be missed." See ibid., xxi.
[68] ibid., 18.   [69] ibid., 19.   [70] ibid.   [71] ibid., 20.   [72] ibid., 21.

He also lectured the forestry department on fire paths. These, he explained, had been invented by Captain Forsyth sixty years previously in the Central Provinces of India. Though Indians imagined the forests "must always burn," and even foresters had once imagined them unstoppable, Forsyth "succeeded in protecting the whole area of demarcated State forests" and succeeded with little cost. Paths were cut and then early in the dry season, burned. This simple measure protected "the whole of the state forests of India." He concluded that "Something like the . . . fire-path . . . might be useful in Australia."[73]

The forestry departments also failed to handle grazing properly – either allowing too much or not collecting needed revenue when it occurred. He scolded officials by reminding foresters that "grazing is not a by-issue: it is an essential part of forest management." He argued for "absolute control" over grazing. Grazing produced not only revenue, but reduced fire risk and, when used correctly, fed animals when grass outside the reserves had been wiped out by drought. "In India, this has become one of the most important drought-relief measures."[74]

South Africa as well, provided examples for Australia where working plans incorporated not only lumbering techniques but also tourism.[75] Australia should hold conservators' conferences to promote "unity of action" and share expertise as in "the early days of Indian forestry."[76] Departmental reports should be yearly. "The yearly reports of the Indian Forest Department have been proposed as models for general adoption . . . The Indian yearly reports are models of administrative ability, carefully sifted figures and condensed details."[77] Though Australia may not be able to afford such detailed reports as India, they should do what they can.

Finally, Hutchins argued that besides improved administration and methodology, the government of Australia must reserve more forest areas. Again the model was India, but also, by 1914, the United States and South Africa, both influenced by the Indian model.[78] Hutchins proposed a more federal approach to lands' policy that would convene under a single forestry service, under a chief conservator with the power to carry out his plans.[79] As the British Empire forestry conferences confirm, *A Discussion of Australian Forestry* was widely read by Australian foresters, and many of the remedies suggested were gradually adopted.

The federal forestry policy recommended by Hutchins came about with the first of the interstate forestry conferences held in Sydney in 1911. The conference saw forestry as "a great national question," where issues required "collective consideration in the interest of the whole Commonwealth" and "national conservancy . . . should provide for the maintenance of a sufficient portion of forest wealth."[80] The second conference in Melbourne in 1912 recommended 10 percent

---

[73] ibid., 25. [74] ibid., 63. [75] ibid., 99. [76] ibid., 144. [77] ibid., 145. [78] ibid., 195.

[79] ibid., 196. The similarities between Indian and even American forest conditions provided many ready-made solutions to technical problems and thus led Hutchins to place a premium on efficient organization. For an early technical discussion of forest conditions in Australia see Peter Macpherson, "Some Causes of the Decay of the Australian Forests," *JRSNSW* 25 (1891): 83–96.

[80] Carron, *History of Forestry in Australia*, 241.

of state lands be reserved, and at the Adelaide conference of 1916, attended by the governor-general, Sir Ronald Munro Ferugson, the Australian *Journal of Forestry* was established as well as a research branch. Since forests covered only 3.3 percent of the land area of Australia, the reservation of forests and the implementation of policy for sustained yield took a backseat to plantations of softwoods and research.[81] But gradually the recommendations of Hutchins were implemented till Australian forestry became integrated with an empire-wide standard of forestry policy and practice.

Though substantial forestry legislation did not occur in New Zealand and Australia until the twentieth century, the debate about the advisability of forestry on the Indian model began in the nineteenth century.[82] The rapidity of forest loss, with vast private domains parceled out to settlers for extensive pastoralism and sheep grazing, did not go unobserved. But the normal colonial experience of settlement, development of farmland, and the growth of cities did not cause such alarm as a rapid immigration for gold, and the subsequent growth of the railroad, opening up forest areas for easy exploitation and in turn requiring sleepers and fuel, and stimulating more demand by spreading settlements inward.

[81] "Timber and Forestry in Western Australia," *Indian Forester* 28 (1902): 85. Edward Scammel, "The Timber Resources of the Australian Commonwealth," *Indian Forester* 28 (1902): 312.

[82] Very little scholarship is available on forestry in Australia and New Zealand in the nineteenth century. A few journals, particularly the *Indian Forester*, did print observations on environmental developments in the region, however. See Macpherson, "Some Causes of the Decay of the Australian Forests," 83–96; "Notice Regarding New South Wales Forestry Department," *JRSNSW* 25 (1891): 30; J Maiden, "Forests Considered in their Relation to Rainfall and the Conservation of Moisture," 211–240; Scammell "The Timber Resources of the Australian Commonwealth," 533; "The Forest Wealth of New South Wales," *JSA* 44 (1896): 51; "Australian Forestry," *JSA* 53 (1905): 144; "Forestry in New South Wales," *JSA* 59 (1911): 363; "Forestry Work in New Zealand," *JSA* 66 (1918): 222; "Forestry in South Australia, 1891–92," *Indian Forester* 19 (1893): 108; A. S., "Annual Progress Report of State Forestry Administration in New South Wales for 1891," *Indian Forester* 19 (1893): 149; "The Condition of Forestry in New Zealand," *Indian Forester* 18 (1892): 416; "Forest Conservancy in Victoria," *Indian Forester* 18 (1892): 436; A. S., "Annual Progress Report of State Forest Administration in New South Wales for 1890," *Indian Forester* 18 (1892): 395; A. A., "Forestry in South Australia," *Indian Forester* 18 (1892): 436; B. Ribbentrop, "The Forest of Victoria," *Indian Forester* 21 (1895): 451; "Forest Administration in South Australia, 1893–1894," *Indian Forester* 21 (1895): 239; J. H. Maiden, "Marram Grass in Australia," *Indian Forester* 21 (1895): 352; "India and Australia," *Indian Forester* 21 (1895): 406; "Forest Work in South Australia, 1895–1897," *Indian Forester* 24 (1898): 169; "The Report of the New South Wales Department of Agriculture and Forestry, for 1895–97," *Indian Forester* 24 (1898): 467; "State Forestry Administration in South Australia," *Indian Forester* 25 (1899): 217; "Forestry in New South Wales," *Indian Forester* 25 (1899): 154; "Forest Administration in South Australia, 1898–99," *Indian Forester* 26 (1900): 99; J. Ednie Brown, "The Forest of Western Australia," *Indian Forester* 26 (1900): 264; "Western Australian Timbers," *Indian Forester* 26 (1900): 154; "State Forest Administration in South Australia for 1899–1900," *Indian Forester* 27 (1901): 256; "Forest Conservation in New South Wales, 1899," *Indian Forester* 27 (1901): 214; "Timber and Forestry in Western Australia," *Indian Forester* 28 (1902): 85; Scammel, "Timber Resources of the Australian Commonwealth," 312; "Australian Forestry," *Indian Forester* 31 (1905): 171; "Eucalyptus Screens as Fire Protection Belts," *Indian Forester* 31 (1905): 297.

In New Zealand, as early as 1877 an article in the *Dundee Advertiser* summarized the influence of Indian forestry in the region. It observed a growing feeling that "scientific forestry in India" had "wisely directed" attention to New Zealand's need for sound forestry policy. Happily, the article said, "There are fewer difficulties to encounter in New Zealand than in India," and the government and legislature "would be well advised to initiate a system of forest conservancy" while enough forests yet remained to conserve.[83]

The article reminded readers that legislators had only recently passed a law allowing for government reservation and had called upon an Indian forester, Captain Campbell-Walker of the Madras forestry staff, to "guide them in laying down plans for making the most of the existing forests." New Zealand foresters from the 1920s saw this period of early activity as a time of "far-seeing parliamentarians, dreaming of things generations before their time." A director of forestry, L. MacIntosh Ellis stated in 1923 that forestry concerns arose from "the virile acorn-seed of staunch English Victorianism . . . planted in the virgin soil of New Zealand social liberalism."[84]

In Parliament, Sir Julius Vogel proved to be one of those far-seeing legislators who pleaded for a forest policy along Indian lines. But Parliament did little until 1873, when Vogel became premier and sponsored reports on the state of the forests to the Parliamentary Committee on Colonial Industries. These reports urged fire control but also clearance and settlement. The most influential report came from Captain Campbell-Walker of the Indian forest service, whose suggestions found their way into the New Zealand Forests Act of 1874.[85]

The parliamentary debate over forests that the *Dundee Advertiser* article mentioned had occurred from 1868 to 1874, ending with the summoning of Campbell-Walker to issue a report. Though no forest department had been established in the 1860s and 1870s in New Zealand, the government collected information on forestry with a view to future conservation of existing forests. A series of these documents was transmitted by the ministry of lands, which in turn helped inform Parliament regarding conservation policy.[86]

The first movement toward conservancy began when a legislator, a Mr. Potts, moved in 1868 "that it is desirable government should take steps to ascertain the present condition of the forest so that the colony [can act] with a view to their better conservation."[87] The question "What timber have we?" had no clear answer, because so little surveying had been done. The only estimate, and a vague one, came from J. Vogel, Superintendent of Forests, claiming approximately 1.5 million

[83] National Agricultural Library, Ministry of Lands, no pagination, manuscript insert into *Papers Relating to State Forests: Their Conservation, Planting, Management, &c.: Presented to Both Houses of the General Assembly* (Wellington, 1874).

[84] L. MacIntosh Ellis, *The Progress of Forestry in New Zealand. Prepared on the Occasion of the Sixteenth Meeting of the Australasian Association for the Advancement of Science* (Wellington, 1922), 6.

[85] ibid., 8–9.   [86] NAL, *Papers Relating to State Forests*.   [87] ibid., 1.

acres available for conservation.[88] Vogel also estimated a decrease from 66 million acres in 1830 to 12 million acres in 1873 of total forest area.[89]

As a result, in August 1874 Parliament passed an act entitled "The Establishment of State forests, and . . . the Application of the Revenues Derivable Therefrom." In its preamble it stated that "it is expedient to make provision for preserving the soil and climate by tree planting, for providing timber for future industrial purposes, for subjecting some portion of the native forest to skilled management and proper control, for these purposes constitute state forests." The act also provided for a method of reservation that would later be used extensively in the United States. This method allowed the governor to reserve forests, not privately owned, by "proclamation," at the discretion of the governor.[90]

Accordingly, Parliament appointed Captain J. Campbell-Walker as Conservator of State Forests in 1876. They asked him to examine the forests of New Zealand and then to recommend measures of conservation to protect climate, forest resources, and soil for generations to come. His report drew directly on Indian forestry expertise and ideas. Rather than immediately setting up a forest department, Julius Vogel suggested that he spend a year touring the Islands and gaining an intimate knowledge of conditions in New Zealand. The India connection could not be clearer: Campbell-Walker stated bluntly that "my services had been lent by the Government of India."

Campbell-Walker's report offered a multiple-use formula borrowed directly from Brandis, seeking to establish "management most likely to prove beneficial, and at the same time least antagonistic to the current public or popular opinion." Campbell-Walker also mentioned "climatic consideration" and water supply along with the need for public revenue. Though no action immediately followed the report, it increased awareness in New Zealand of the importance of forest conservation for the protection of not only the timber supply, but also for water, soil and climate considerations.[91]

The report proved instructive. While advocating forest reservations to protect against declining rainfall, it insisted a forest department could be revenue-producing like that of India, if the colony only acted "to reserve absolutely a

[88] ibid., 26.   [89] ibid., 35.

[90] A. Lecoy, *Suggestions on Forests in New Zealand* (Wellington, 1880), 2, 3, 8; Kirkwood, *Papers and Reports Upon Forestry*, 48.

[91] Campbell-Walker, *Report of the Conservator of State Forests*, 1–2, 53; NAL, *Papers Relating to State Forests*, part 2, 3. Joshua Hooker, director of the Kew Gardens, whose influence had so heavily impacted the development of forestry in India, also influenced the New Zealand parliament, as can be seen by his letters to various New Zealand officials that were found in the packet of state papers requested by the colonial parliament. Hooker preached "the duty of conserving national resources of the colonies for the benefit of future generations, whilst encouraging a fair use of them by the present." NAL, *Papers Relating to State Forests*, part 2, 8, 13. By the 1870s the link between deforestation and climate change became international in scope and not limited to the colonies. See W. Christie, "The Forest Vegetation of Central and Northern New England in Connection with Geological Influence," *JSNSW* 11 (1878): 21; Clarke, "Effects of Forests on Climate," 179; Maiden, "Forests Considered in their Relation to Rainfall and the Conservation of Moisture," 211–240.

small proportion of the un-alienated forest area." It concluded with six concrete suggestions for setting up a successful department: (1) the formation of a state forest department for colonial forests, "to be set apart for climatic considerations and the permanent supply of timber," (2) an act empowering the governor to appoint officers, declare reserves, and regulate crown property, (3) the selection, demarcation and survey of reserves for "systematic management" by departmental officers, (4) the assessment and assignment of remaining forest area to the department or for private sale, (5) the formation of plantations of timber, and (6) the encouragement of private planting.[92]

Interestingly, this early phase of forest reservation did not prioritize revenue or the collection of royalties from timber. A report of 1880 to Parliament by the minister of lands, A. Lecoy, stressed that the reserves, in addition to timber, were to "maintain the protection given by Nature against the disturbance of the climatic equilibrium" in addition to droughts and flood control. France, which offered such excellent training to foresters, served as a bad example for New Zealand. "In that country," Lecoy cautions, "as a result of injudicious alienations of State forests and the further conversion of the forest land into pasturages" the people are now subject to "periodical inundations." They face the ruin of agriculture in numerous districts and a depopulation of the countryside, with massive immigration to America. Therefore, he concluded, the reservation of forest lands cannot be for timber alone, but for the "general interest."[93]

Early foresters in New Zealand advocated the reservation of forests for the protection of soil and climate. Gradually, under the guidance of a trained botanist, Professor Thomas Kirk, appointed in 1885 as chief conservator, officials laid down rules for scientific forestry. These included regulation for the harvesting of timber, demarcation, fire prevention, and fines for illegal removal of timber. Felling should be by rotation, with a ban on the felling of trees in mountain areas that protect soil and prevent floods. Over 800,000 acres were set aside in Kirk's term. By 1887 he felt confident that "the enormous waste and robbery" of the past had been greatly reduced.[94]

But public interest in forestry and the preservation of nature rose and fell. In 1888 the government curtailed application of the Forests Act and dismissed almost all the forest officers, even Kirk. Attempts were made to change the direction. In 1896 a timber conference issued a report to Parliament, followed by an impassioned plea by George Perrin, Conservator of State Forests in Victoria, for full application of the Forests Act and further reservation, along with scientific forest management. But parliamentary reports did not change the facts on the ground, and the movement stalled for another decade or more.

By 1909 the American conservation movement, itself modeled on empire forestry, helped awaken a response in New Zealand, as did the reservation of forest areas and intense management in the southern Australian states. Some officials

[92] NAL, *Papers Relating to State Forests.*   [93] Lecoy, *Suggestions on Forests in New Zealand*, 1.
[94] Ellis, *Progress of Forestry*, 9.

voiced concern that the state forests of New Zealand were, without proper management, merely reserves for future sawmills. Still the government took little action until 1913. Reporting to Parliament, a royal commission on forestry urged that a forest department be created, with a forestry code, tight administrative control, and the planting of plantations to relieve the strain on indigenous forests and meet timber needs. It also gave many reasons for conservation that went beyond the normal arguments of timber supply and resource allocation. Reserves were recommended for the protection of historic places, scenery, hot springs, soil, wildlife, and flora. Echoing Ribbentrop's "household of nature" argument, the commission conceived forests as "open-air museums" where plants and animals would be left unmolested for future generations. In addition, the commission argued, nature should be set aside for picnics and other forms of recreation, increasing popular support.

All these ideas showed a deep familiarity and variation on the concept of multi-use forestry and point toward the trend of moving from minimum to maximum protection of forest lands, a trend characterized by the conservation movement at large over the last century and a half. Turning land from timber reservation to "sanctuaries for the flora and fauna" where "no firearms may be discharged . . . nor may any bird or game be killed" is a clear line of progression to fuller protection.[95]

The commission also reported to the minister of lands the arguments made by those who wished to reverse the process and downgrade "certain of the scenic reserves" from protection. Some suggested that much of the land was suitable for farmland, and necessary for further settlements. Others that the reserves harbored "noxious weeds" which spread to neighboring farmlands. Others that the reserves had suffered a forest fire and no longer served recreational purposes; few picnic on the land anyway and, finally, reserves will likely burn by accidental fire again in the future.[96]

Soon afterwards came a visit by D. E. Hutchins, who issued a report that resulted in profound changes. The respect with which forestry officials received and introduced Hutchins to Parliament gives an indication of how influential his recommendations proved. F. H. D. Bell, the leading forestry official in New Zealand, described him as an Indian forester with a specialty in tropical forestry, who transferred to the Cape for twenty-three years to spread "a love of forests and an interest in arboriculture in that almost treeless country." Due to his afforestation efforts, he wrote, "from Cape Point to the Limpopo River, may be seen extensive plantations of the Australian eucalypts." Hutchins further organized a forest department in

[95] Aiding in the implementation of the Scenery Preservation Act of 1908, and a keen observer of developments in Australia and the United States, W. C. Kensington, Under-Secretary of Land, issued *New Zealand Department of Lands Report on Scenery Preservation for the Year 1908–1909* (Wellington, 1909); see 3, 85. See also W. C. Kensington, New Zealand Department of Lands Forestry in New Zealand (Wellington, 1909); *Royal Commission on Forestry*, (Wellington, 1909), xvi–xvii; Ellis, *Progress of Forestry*, 10.

[96] *Royal Commission*, xvii.

British East Africa and issued a seminal report on forestry in Cyprus, and issued a report on forestry for Australia.[97]

Thus his previous experience with forests "in India, Africa and Australia" enabled him to bring knowledge and experience "in excess of those possessed by any previous investigator," and his reports, Bell concluded, will require the "careful attention" of all who value "the perpetuation and protection of our native forests." A recent report by Hutchins had also been launched in Britain, he added, which at long last was beginning to initiate forestry reserves and plantations in sync with the colonies. With such wide expertise and influence, New Zealand was fortunate, he enthused, to have Hutchin's advice.[98]

What did Hutchins say that provoked such enthusiasm? He mixed generous doses of romantic appeal with utilitarian arguments. It is true, as with most empire foresters, that a multiuse revenue-producing forest persuaded the public and legislators more than romantic appeals. But along with this hardheaded approach Hutchins employed powerful poetic imagery. He quoted at length from Mr. T. F. Cheesman's "Illustrations to the New Zealand Flora":

From a considerable distance ... are at once recognized ... the adjoining forest, ... by the dusky-green colour of the foliage. But it is from the interior of the forest that the Kauri is seen to the best advantage, and the majestic size and noble proportions of the tree can be best appreciated. On all sides rise the huge columnar trunks, sometimes towering up for more that 80 feet without a branch, and tapering but slightly from base to summit, smooth, grey, and glistening. At the base of the trunk is the huge mound of debris produced by the fall of the bark, which is regularly cast off in large flakes ... From the top of the trunk spring the immensely thick branches, often growing out almost from a single point. These, with the branches and foliage, form a high-vaulted roof to the forest, through which a varying amount of daylight filters through to the ground ... Under the vault or roof of branches the eye can penetrate far and wide among the massive trunks, which have hence been compared to the pillars of some gothic cathedral.[99]

Hutchins had the gift of mixing scientific observation with a careful analysis of local conditions and needs, packaged in impassioned propaganda. He hinted that patriotism required the saving of the great softwood forest of New Zealand, the Waipoua Forest. He reviewed local history, pointing out the eminent men who admired the great kauri tree, emperor of the forests: Caption Cook, who discovered the useful resin of the tree; Sir Rider Haggard the novelist, who wanted to visit but could not, due to the remoteness of the forest; Darwin, whom he quotes from his *Voyage of a Naturalist*, "the noble kauri trees, the most valuable production of the Island."[100]

Hutchins certainly pleaded financial considerations as well. The largest tree ever found in the United States, "the mother of the forest" in Alaveras Grove, had 140,619 board feet, a huge mass of timber for one tree. But even this the largest

[97] D. E. Hutchins, *Waipoua Kauri Forest, its Demarcation and Management* (Wellington, 1918), 5–6.
[98] ibid.   [99] ibid., 7.   [100] ibid., 1–7.

kauri tree dwarfed, for the California tree "contained less than half the timber of the recorded largest kauri of New Zealand." This was not a difficult to sell hardwood, which the world used in only 10 percent of its market, but a valuable softwood needed for an ever-growing consumer demand. Hutchins appealed to national pride, by pointing out that these valuable forests stood "supreme" in their "own class." The woods were better than "the virgin forest of South Africa, better than that in the same forest on the highlands of British East Africa, and much better than that on the Nilgiri Mountains of India." Yet, he scolded, though New Zealand has the best forests, in all these other countries this class of forest "is being preserved by the governments concerned."[101]

The war provided useful lessons as well. The Australian states had already begun, after his suggestion, an employment plan for returning soldiers to work and preserve the eucalypts, a project reported in the *London Times* of July 10, 1918. Why not the same for New Zealand? This could attract immigrants from Britain, who were all medically fit (for the army). Additionally, this populated rural areas and fought national decay. This last remark is particularly interesting, since it shows Hutchins felt that urban life led to cultural and national decadence, and along the lines of the Romantics in England and Germany, that national renewal lay in a balance between urban and rural areas.[102]

This last reason for conservation put Hutchins firmly in the political philosophical camp of the Bull Moose Party in the United States, the Tories in Britain, and in a few short years, the National Socialist in Germany. It was a belief that came to be widely employed by environmental propagandists up to the Second World War, and to a lesser extent afterward. Hutchins knew the power of the argument of rural–urban balance and used it skillfully by including in his report war maps of the French and German sides of the Rhine valley. He concluded that this highly industrialized area, home to one of the world's highest concentrations of industries, including the great Essen works, dye-works, and chemical works, nonetheless had 33 percent forest cover. Clearly, conservation can work with development and progress, he reasoned.[103]

He concluded his appeal for the reservation of the kauri forests and the establishment of a forest department along Indian lines by quoting Tennyson,

> My father left a park to me, but it is wild and barren.
> . . .
> Yet, say the neighbors when they call, this is not bad but good land,
> And bears in it the germ of all that grows within the woodland.[104]

The government took action and, under Bell, commissioner of state forests, introduced his report to Parliament. This inaugurated a forestry program that finally began to emulate the Indian exemplar, with the objectives of (1) providing timber supplies, (2) providing protection against flooding and desiccation, (3) protecting

---

[101] ibid., 178, 184.   [102] ibid., 184.   [103] ibid., 188.   [104] ibid., 189.

existing forests (4), and protecting forests for future generations, including protecting forests for water, soil, climate, and public health. These goals resulted in substantial afforestation, the suspension of *laissez-faire* principles on state property, the assertion of absolute government ownership of nonprivate land, and the encouragement of community forestry. To these accomplishments were added the protection of "native birdlife" and education of the public of the necessity of conservation.[105]

Canada

The empire forestry model worked in Canada for the same reason the model worked for its southern neighbor – it provided a revenue-producing multiuse forestry that appealed to economics as well as "the sentimental aspect."[106] Though Canada had served as a model for the United States on forest fire protection, Canada formed its forestry service in the provinces as late as 1910.[107] In 1886 J. H. Morgan, Forestry Commissioner of Canada, issued a *Report on the Forests of Canada* calling for implementation of legislation placing Canada's forest wealth in reservation. The Minister of the Interior, D. L. Macpherson, asked Morgan to issue the report, instructing him to prepare a paper for the governor-general. The report revealed a concern for concerted action between the Dominion government and the provinces, and revealed the attitudes and inspiration behind the conservation movement in Canada at the time.[108] India loomed large in the report.[109] Besides the call for a halt to forest fires and destructive lumbering practices, and a recommendation that a commission be appointed to propose legislation, the report looked closely at India and other colonies that had adopted sound forest policies.

As a result, the Royal Commission on Forests, Reservations, and National Parks convened to tackle the issues raised. In 1887 Prime Minister Macdonald appointed Morgan as "Forest Commissioner," a position he held through 1889. While giving broad support for the development of conservation in the provinces, he also proved instrumental in passing the Dominion Lands Act (1884), the Dominion government's first step toward a Canada-wide forest reserve system, and the

---

[105] Ellis, *Progress of Forestry*, 10–13.   [106] *British Empire Forestry Conference*, 77.

[107] A few fragments of Canadian forestry history can be gleaned from the following articles, "Forest Wealth of British Columbia," *JSA* 47 (1898); "The Perpetuation of Canadian Forests," *Indian Forester* 26 (1900): 120; Harold Unwin "Canadian Forests and Forestry," *Indian Forester* 52 (1904); "Forestry in Canada," *Indian Forester* 31 (1905): 667; "Canadian Forestry Journal, for March 1907" *Indian Forester* 33 (1907): 465.

[108] R. Peter Gillis and Thomas Roach, *Lost Initiatives: Canada's Forest Industries, Forest Policy and Forest Conservation* (Westport, CT, 1986), 44.

[109] J. H. Morgan, *Report on the Forests of Canada, in which is shown the Pressing Necessity which Exists for their more Careful Preservation and Extension by Planting, as a Sure and Valuable Source of National Wealth* (Ottawa, 1896), 5.

"first flowering of forestry in Canada."[110] As two environmental historians have noted, "From this combination of forest preservation and land and general watershed conservation, federal forest policy was to grow."[111]

After the American Forestry Congress convened in Montreal, the Ontario government appointed a "clerk of forestry" to promote conservation, what we would today call a publicist.[112] A former journalist, the new "forestry publicist" Robert W. Phipps issued in 1883 a *Report on the Necessity of Preserving and Replanting Forests*.[113] He followed a well-known course of argument, "generally observed in other countries," by beginning with the scientific link between forestry and climate, and "some very useful and exhaustive reports concerning the examinations made by the East Indian Government in the system of European Forestry" brought to his attention by Canadian forestry advocates M. Joly and Goldwin Smith.[114]

He also drew on the "reports of forests in Australia, New Zealand and India," and not least, the United States, particularly the work of "Prof. Hough," who also responded with enthusiasm to the Indian forestry model.[115] By 1883 the web of influence had become sufficiently diffuse for Phipps to mention a well-known chronology: German forestry methods put to use and expanded in India, along with new legislative concepts, and adopted by the colonies and the United States. In this report Phipps detailed the developments that flowed from the Forest Charter and the reservation of wasteland into absolute government property. He also stressed the need to emulate India, South Australia, New Zealand, and the United States, exactly retracing the steps of influence from India, the colonies, and then the United States.[116]

An influential successor to the office, Thomas Southworth, wrote a memorandum to J. M. Gibson, Commissioner of Crown Lands, calling for a commission to investigate the "whole question of forest management in Ontario."[117] Gibson agreed and ordered a report of the Royal Commission. Alexander Kirkwood, Chief Officer of the Lands Branch of the Department of Crown Lands, issued a supplement in 1893 on forestry methods from around the world as a guide for Canadian policy. Kirkwood, an Ontario civil servant, played a key role in the formation of Algonquin Provincial Park, an area that preserved forest and wildlife.[118] He drew heavily on the work of W. Schlich, former inspector general of forests in India, and included a region-by-region analysis of progress in the British colonies, with Indian forestry especially in mind. In a search for a "model forest," the author turned to India. He stated bluntly that "Practically, only India has really

[110] Gillis and Roach, *Lost Initiatives*, 46.   [111] ibid.   [112] ibid., 42–43.

[113] Robert W. Phipps, *Report on the Necessity of Preserving and Replanting Forests: Compiled at the Instance of the Government of Ontario* (Toronto, 1883).

[114] Phipps, *Report*, 3.   [115] ibid.   [116] Phipps, *Report*, 95–98.

[117] J. M. Gibson, *Report of the Royal Commission on Forestry Protection in Ontario* (Toronto, 1899), 5–6.

[118] R. Peter Gillis and Thomas Roach, "Early European and North American Forestry in Canada: the Ontario Example, 1890–1941," in *History of Sustained-Yield Forestry: A Symposium*, ed. Harold K. Steen (1984), 215.

and honestly dealt with the forest question."[119] After a history of forestry policy in India, the most detailed of any region the report covers, Kirkwood concluded that:

In this manner a well-organized department has been built up [in India] during the last quarter of a century, which has under its charge an immense government property consisting at present of some 55,000,000 acres of forest lands. Some of the forests were taken in hand before they had been destroyed, but by far the greater part of the area was taken over in a reduced and even ruined condition. Although a quarter of a century is only a short period in the life of a timber tree, the effects of protection and systematic management are everywhere apparent. Economic systems of utilization have been introduced, a larger portion of the forest is successfully protected against the formerly annually recurring forest fires, young growth is allowed to spring up under the protection afforded; sowing and planting are carried out when required; the forests are managed under carefully considered working plans; and all this without interfering with the acknowledged rights of the people, who receive every year enormous quantities of forest produce, whether free of charge or at comparatively low rates. In many parts of the country the people have come to recognize the importance to themselves of the proper preservation of a suitable forest area.[120]

The report of the 1897–99 Royal Commission persuaded the provincial Ontario government to reserve large forest areas, covering thousands of square miles, for environmental management and fire protection.[121] This advance was "a major advance in forest conservation when measured by any yardstick."[122]

Unlike the United States, no towering figure can be traced in Canadian forest history that single-handedly spearheaded the drive for forest protection. The introduction of empire forestry method to Canada occurred over many decades and did not mature until after the United States initiated its own forest service, itself modeling on the Indian exemplar. Though later criticized for shortcomings that fell beneath the high professional standard of forestry in India, the 1920 Empire Forest Conference makes clear that Canada had joined, incrementally, the imperial forestry team modeled on the innovations in India.[123]

The rest of the empire

With no forestry program in Great Britain, India served as an environmental center, with a fund of trained expertise that the Indian Forestry Department exported throughout the British colonies and to over one-third of the land surface of the earth.[124] Interest in forestry, and thus in the Indian experiment, increased in the

---

[119] Kirkwood, *Papers and Reports Upon Forestry*, 52.   [120] ibid., 59.

[121] Gillis and Roach, "Early Forestry in Canada," 216.   [122] ibid.

[123] A. Koroleff, "Fundamental Cause of our Failure in Forestry and the Remedy," in *Report for the Forestry Research Conference at the National Research Council of Canada* (Ottowa, 1935), printed in *Woodlands Review* 6 (1935).

[124] For an analysis of how India influenced the teaching of forestry in other colonies see Pearson, "Teaching of Forestry," 422–478.

latter part of the nineteenth century. One spur proved to be the International Forestry Exhibition in Edinburgh in 1884, which exhibited the woodland products of the world, including South America and Japan, followed by a colonial and Indian exhibition.[125] "Good writing and speaking on the subject" inundated the Colonial Institute, the Imperial Institute, and various magazines.[126] The forest school established at the Royal Engineering College at Cooper's Hill began to train foresters "not destined for India," who would serve in the other colonies and prove pivotal in initiating forestry programs in the colonies.[127]

General J. Michael, who served in India and was an active contributor to the *Journal of the Society of Arts*, exuberantly credited India for "her marvelous successes during the last half-century," and for having "set in motion and stimulated the recent tide of public interest in forest conservancy throughout the colonies."[128] He reeled off the names of the colonies who derived their forestry programs from India: Ceylon, New South Wales, New Zealand, the Cape, Mauritius, Jamaica, and Cyprus. They "have all borrowed officers from India to put them in the way of organizing conservancy and the economic working of their forests." Even the United States had "instituted a division of forestry in their department of agriculture." Indian foresters constituted the current heads of the forest departments in most of the colonies, including such entities as the forestry school at Cooper's Hill, the Lecturer on Forestry at Edinburgh University, the editor of the *Indian Forester* and other forestry journals, the president of the Royal Scottish Arboriculture Society, and other organizations too numerous to name.[129] Indian forestry had done more than leave its mark; it had initiated empire forestry and the first worldwide environmental program.

The influential British association often drove the need for forestry home to the scientific elite of their respective countries. For instance, in a report on the commercial aspects of forestry in Australia, E. T. Scammel, a forestry official, argued that the self-governing colonies ought to "consider the magnificent example which has been set them by India, where the preservation of the state forests has now been put on a safe basis, for the everlasting benefit of the people and the country." He argued that it is incumbent upon "British Colonial Governments to give the question of forest control and development their most careful and enlightened consideration." Australia, he argued, should not care less about its future than India.[130]

The empire had forestry protection for the vast majority of its domain. When Indian forestry became empire forestry, it covered in the British territory alone almost half the forests in the world. Add to that the administration of forests of the United States and other countries like China, who imported Indian expertise, and a truly world picture of early environmentalism emerges.

Though this investigation has focused on the larger self-governing dominions, a host of other territories also adopted the forestry practices pioneered in India. To

---

[125] Michael, "Forestry," 94.   [126] ibid.   [127] ibid., 96.   [128] ibid., 102.
[129] ibid., 103. See also Schlich, "Forestry in the Colonies and in India," 192.
[130] "The British Association at Southport," *Indian Forester* 30 (1904): 35.

reconstruct the forest history of every territory of the empire in one book would be impossible, but one example, Cyprus, illustrates the concern colonial administrators exhibited in organizing forestry within the scope of the entire empire.

A treaty with the Sultan of Turkey in 1878 gave the British administrative rule over Cyprus, the third largest island in the Mediterranean. With the outbreak of war in 1914, Britain annexed Cyprus outright. With a population of only 347,000 in 1931, Cyprus boasted a typical warm Mediterranean clime and a very long and dry summer, with little rainfall in the winter. This naturally placed a premium on water and – in the minds of empire foresters – on the trees of the island.

Its forests grew largely in the mountains, with various species of pine, cedar, and cypress, but also some hardwoods like oak and juniper.[131] Forest had originally covered the whole island, but in the Middle Ages shifting cultivation took its toll even as exports dropped, with the Turkish occupation and settlements hastening the destruction of forest. In 1870, to conserve trees, the Turkish authorities banned timber exports, and in 1873 the Ottoman regime hired a French forester, M. de Montrichard, to investigate the cause of the timber shortage and to recommend a course of action. But his suggestions went unheeded, and it was not until British occupation in 1878 that the forests' destruction ended and the authorities initiated a program to reverse the decline.

Various writers, British as well as French, expressed outrage at the waste of forests. Samuel Baker in *Cypress As I Saw It In 1879* wrote:

There is no sight so exasperating as this uncalled-for destruction; it is beyond all belief, and when the amount of labor is considered that must have been expended in this indiscriminate attack upon forest trees that are left to rot upon the ground where they have fallen, the object of the attack is at first inconceivable. The sight of a mountain pine forest in Cyprus would convey the impression that an enemy who had conquered the country had determined utterly to destroy it, even to the primeval forests; he therefore felled, and left to rot, the greater portion of the trees, but finding the labour beyond his means, he had contented himself with barking, ringing and hacking at the base of the remainder, to ensure their ultimate destruction.[132]

In response the British dispatched an Indian forest officer, A. E. Wild, to tour the island and report on the condition of the forests. Dismayed at the exploitation of whole forests to obtain only minor forest product, Wild advocated a forest law to reserve large tracts of land and initiate demarcation and working plans.[133] As a result of his recommendations, Forest Law no. 22 of 1879 placed all forest area other than private and corporate property under the control of the state.[134]

Hutchins also issued a *Report on Cyprus Forestry* in 1909, which served as an inspection tour to realign closer to empire forestry standards the forest department

[131] Troup, *Colonial Forest Administration*, 396–397.
[132] ibid., 397. Other observers were equally appalled. See M. Bricogne, "Les forêts de l'empire Ottoman," *Revue des Eaux et Forêts* 16 (Nancy, 1877): 471.
[133] A. E. Wild, *Report on the Forest in the South and West of the Island of Cyprus* (London, 1879).
[134] Troup, *Colonial Forest Administration*, 401.

that had been created in response to Wild's report. He wrote that when the British took control of Cyprus from Turkish rule, "the state of waste and destruction that existed then was reducing the forested area dramatically . . . When the British Government took over the administration of the land of Cyprus, the evil of goat-grazing and the necessity of restoring the forests was admitted on every side." The goats that swarmed the island had to be controlled, hence forest laws were enacted that regulated grazing, as in India.[135] Hutchins wrote that "forest laws of Cyprus, closely following Indian legislation, are excellent."[136]

The British encountered special problems in the nearby territory of Palestine. Under the stress of Zionist agitation and terrorism, the forest department nonetheless had set aside 385 square miles as forest land with 276 reserves. A recommendation for further afforestation submitted to the Palestine Royal Commission in 1937 by Joseph Weitz, forestry officer of the Jewish National Fund, suggested the government expropriate almost 800 square miles of land for reforestation. But this last suggestion the Royal Commission rejected on the grounds that the plan required massive dislocation of Palestinians from prime agricultural land. Empire forestry did involve circumscribing the rights of forest access for environmental purposes, but not the wholesale dislocation of indigenous people.

Hutchins noted that a forest department in Cyprus had been in place for thirteen years. But he credited Winston Churchill, who toured the island, for giving the forest department the added authority and funds it needed to carry out its work.[137] Hutchins noted that as under-secretary of state for the colonies, Churchill took a keen interest in reforestation. His imperialist bent perhaps inclined him to take a long-term view of the forest area under British control, viewing environmental policy as profitable and identical with holding the empire together.[138] Churchill had recommended, as with British East Africa, the demarcation of forest areas and the organization of the department along the lines drawn up in the Cape and in India.[139]

Subsequently the work on Cyprus served as a model for much of the forest work in the Mediterranean region, particularly Palestine, where denuded hills and the high grazing of goats and sheep caused massive soil erosion. In this territory, under British control after the First World War, a Woods and Forests Ordinance was passed and in 1929 a forest service inaugurated. By 1936 in a land area of 10,400 square miles, 3,860 were desert with only 6,240 square miles capable of supporting either agriculture or forest.

---

[135] Ordinance 22 of 1879, Article 25, Law of VII, of 1881, Article 2. See D. E. Hutchins, *Report on Cyprus Forestry*, (London, 1909), 24.

[136] ibid., 53.   [137] Troup, *Colonial Forest Administration*, 399.

[138] Hutchins, *Report on Cyprus Forestry*, 13. The Liberal Party did not advocate forestry protection as did the Tory Party, because Liberals were concerned that the growing conservation movement compromised property rights – which it certainly did. The Tory Party showed a clearer interest in the environmental protection both at home and in the colonies. See "Afforestation in England," *Indian Forester* 20 (1894): 73.

[139] Hutchins, *Report on Cyprus Forestry*, 14.

Many other countries far from the self-governing Dominions came under the care of empire forestry. In Malaya, the organization of a forestry department occurred when H. N. Ridley became Director of Gardens in 1888. The annual report of the forests administration in Malay records in 1935 that to him "we owe the existing forests . . . for his foresight in causing them to be reserved." Ridley recommended an organized forest department "with a trained officer from India at its head," but no action was taken until in 1900 H. C. Hill, "of the Indian Forests Service," was commissioned to make recommendations for the management of the forests. Their chief forest officer, A. M. Burn-Murdoch, transferred from Burma in 1901 and organized the department along classic Indian lines. Outside the self-governing Dominions, the Malayan department proved second only to the Nigerian Forest Service and had, by 1937, reserved 10,471 square miles, approximately 20 percent of the total land area of the Malay states.[140]

By 1895 Hong Kong had an active Botanical and Afforestation Department,[141] British Guiana had an active Institute of Mines and Forests[142] as did the West Indies and the Sudan, when it came under British control.[143] Indian forestry expertise served not only the British Empire but also other countries interested in establishing forestry departments. China established a forest department in 1916, advised by an empire forester, F. Sherfesee, under the Ministry of Agriculture and Commerce. Another forester, trained at Kew, William Purdom, served as chief of one of six forestry divisions in China.[144] And the United States, as will be examined in the next chapter, adopted Indian forestry innovations and set up, outside the forty-nine states, an American Bureau of Forestry that operated in the Philippines in 1904 and in Hawaii in 1905.

[140] Troup, *Colonial Forest Administration*, 375, 376, 382.
[141] "Afforestation in Hong Kong," *Indian Forester* 31 (1895): 350.
[142] "Forest Products of British Guiana," 99.
[143] "Forestry in the Soudan," *Indian Forester* 29 (1900): 367; Muriel, "Forest Exploration in the Bahr-el-Ghazal (Sudan)," 401.
[144] "China," *JSA* 66 (1918): 687.

6

Empire forestry and American environmentalism

In the nineteenth century the United States government transferred 1 billion acres of public land into private hands, one-half of the land mass of the continental United States.[1] The Department of the Interior deemed public land either suitable for agriculture or not, with forest areas devoid of special designation. Railroad companies received large grants of land, as well as state-sponsored universities (today known as land grant schools), while speculators and settlers purchased or claimed land for the westward migration. Land could be purchased cheaply, and Congress divested the federal government of land as quickly as the market would absorb it. In spite of this great divestiture, the surprising fact is that by the First World War a large section of forests remained in the public trust, managed by a professional cadre of government foresters.[2] Three central foresters, Franklin B. Hough, Charles Sargent, and Gifford Pinchot, credited empire forestry, particularly as practiced in India, with

---

[1] Early American land policy envisioned little federal ownership of land. The Land Ordinance of 1785 lowered prices on federal lands and encouraged settlement by farmers willing to work 160 acres. See Ordinance May 20, 1785, 28 J. Continental Congress 375. In order to encourage a stable society of landowner and small gentry, the transfer of land by allocation was favored over cash in the following acts: Preemption Act 1830, ch. 208, 4 Stat. 420; Homestead Act 1862, ch. 75, 12 Stat. 392, 393; Timber Culture Act 1873, ch. 277. 17 Stat. 605; Timber and Stone Act 1878, ch. 151, 20 Stat. 89.

[2] In 1881 Congress established the Division of Forestry under the Department of Agriculture, though its advisory role gave it little authority. But the General Revision Act of 1891 contained what is often called the Forest Reserve Act and authorized the President to create, by proclamation alone, new forest reserves. In 1905 the Bureau of Forestry became the US Forest Service. See the General Revision Act, ch. 561, 26 Stat. 1095. Section 24 reads: "That the President of the United States may, from time to time, set apart and reserve, in any State or Territory having public land bearing forests, in any part of the public lands wholly or in part covered with timber or undergrowth, whether of commercial value or not, as public reservations, and the President shall, by public proclamation, declare the establishment of such reservations and the limits thereof." The General Revision Act is paralleled by expansion in the economic sphere with, in 1887, the Interstate Commerce Act, ch. 104, 24 Stat. 379 and in 1890, the Sherman Act, ch. 647, 26 Stat. 209. To place conservation legislation in a broader political context, see Hays, *Conservation and the Gospel of Efficiency*. Also helpful is Elmo R. Richardson, *The Politics of Conservation: Crusades and Controversies, 1897–1913* (Berkeley, 1962).

the political compromise that led to massive forest reservations by Congress, and the beginning of modern environmental practice in the United States.[3]

The British colonial example ranked as an unparalleled antecedent to the environmental problems faced by Americans, and featured a highly articulate philosophy of management with a powerful ratiocination for public ownership. The presence of a revenue-producing paragon advanced conservation efforts precisely at the moment when the federal government precipitated either massive forest reservations or a final disposal of forest area to private companies.[4] Two-thirds of the federal land mass had been transferred to settlers, institutions and, in some cases, to the state governments. But 500 million acres remained in federal hands in 1890 and, coupled with the perception that the frontier had closed, pressure groups closed in to advocate a final reckoning for the remaining land. The US Census in 1890 proclaimed: "Up to and including 1880 the country had a frontier of settlement, but at present the unsettled area had been so broken into by isolated bodies of settlement that there can hardly be said to be a frontier line. In the discussion of its extent and its westward movement it can not, therefore, any longer have a place in the census reports." Conservation of millions of acres of forest lands for timber, water supply, wildlife, forage for cattle, and recreation all commence in this period.[5]

Franklin B. Hough and empire forestry

Franklin B. Hough served as the nation's first Federal Forest Agent, and the division of forestry owes its origin largely to him. After a stint with the Sanitary Commission during the civil war, he served as Superintendent of the 1870 US census. While compiling timber data for the project, he grew alarmed at the rapid depletion of forest resources and wrote a paper on forestry for the American Association for the Advancement of Science. In 1876 Franklin Watts, Commissioner of Agriculture, appointed him as the nation's first forest agent. In this position he wrote his monumental *Report on Forestry*, published over the tenure of his office, the most comprehensive account of the condition of the US public forests at the time.[6]

---

[3] Michael Williams provides the best overall background and context to the history of forests and forest conservation in the United States in *Americans and their Forests: a Historical Geography* (Cambridge, 1989).

[4] See Bureau of the Census, US Department of the Interior, Compendium of the Eleventh Census, 1890, part 1 at 48 (1892). Homesteaders successfully placed provisions in both the Republican and Democratic platforms between 1872 and 1888 to privatize the remaining land in small plots. A variety of lobby interests pressured Congress in the 1870s and 1880s to relinquish control of the remaining acreage. See Gary D. Libecap, "Bureaucratic Opposition to the Assignment of Property Rights: Overgrazing on the Western Range," *Journal of Economic History* 41 (1981): 51.

[5] Harold Steen, *The US Forest Service: A History* (Seattle, 1976), vi.

[6] "Franklin B. Hough," *American Forests and Forestry Life* (July 1922). "Remembering Franklin B. Hough," *American Forests* 86 (1977): 34–37, 52–53.

With an active interest in geography, physics, meteorology, and botany, Hough served as America's first Chief of Forestry. He put his prodigious talent for statistics to work for the 1870 census of the United States, where he noticed a decline in forest land and brought the decline to the attention of the United States Congress. From a reading of Marsh's *Man and Nature*, he accepted the view that deforestation led to dramatic climate changes. In 1873 he gave an address to the American Association for the Advancement of Science (AAAS), entitled "On the Duty of Governments in the Preservation of Forests." Here he argued for a system of training and forestry management similar to the forestry policies pioneered in the British colonies, especially British India. On the basis of this report, the AAAS memorialized Congress to appoint a chief forester to report on the condition of the nation's forests. The memorial, written primarily by Hough, is entitled *Cultivation of Timber and the Preservation of Forests* and heavily engaged the example of British India, as did his earlier report to the AAAS. The memorial produced an invitation for Hough to meet with President Ulysses S. Grant, at which time he, along with George S. Emerson, a Harvard botanist, discussed "for some time [issues] about forestry."[7] After gaining the approval of the Secretary of the Interior, Columbus Delano, President Grant forwarded the memorial to Congress, who in 1876 approved the bill and voted $2,000 to pay the salary of the United States' first forest agent, under the Department of Agriculture. Agriculture Secretary Franklin Watts chose Hough for the position, which he held until 1883.

In his memorial to Congress Hough discussed an impending timber shortage in the US and the need for an American Evelyn. He also pointed out the effects of deforestation on climate and the need to plant trees to fight flooding. But forestry practice in India drew his special attention. He quoted extensively from Captain Ian Campbell-Walker, Deputy Conservator of Forests in Madras, who calculated the needs of a 200 million population for building material and firewood. Hough observed that the government of India had to "consider such questions as climate, rain-fall affecting the irrigation and cultivation of thousands of acres, and supply of wood-fuel to the railways."[8] The responsibility of all this "devolves on the government" where the necessity of preserving forests involved, in Hough's words, an "equilibrium of temperature and humidity." Forestry in British India implicated the "social welfare" of the populace as well as the welfare of industry and railroads, arts, and "daily utility."[9]

Hough also sketched the history of forestry in British India. The British had "laid the foundation of an improved general system of forest administration" by conserving state forests, and developing state forests as state wealth. That "all superior

[7] Franklin B. Hough, *Diaries*, Nov. 20, 1873, Feb. 2–12, 1874, Franklin B. Hough Papers, New York State Library, Albany.
[8] Campbell-Walker, as quoted in Franklin B. Hough, *On the Duty of Governments in the Preservation of Forests* (Salem, 1873), 48. Taken from Campbell-Walker's *Reports on Forest Management in the Madras Presidency* (Madras, 1873).
[9] ibid.

government forests are reserved and made inalienable, and their boundaries marked out to distinguish them from waste lands made available for the public," were principles worthy of American emulation. The Indian Forest Act of 1865 defined the "nature of forest rules and penalties" and the "executive arrangements" of the local administrations, while surveys also "obtain accurate data concerning the geographical and botanical characterizations of the reserved tracts."[10]

Perhaps most impressive of all to Hough, the Indian government had created a forestry educational system where none previously existed. By sending officers to the European forestry schools, a professional cadre of officers was created. Hough quoted Dietrich Brandis as saying that "with great perseverance and industry these officers went through a regular course of studies in the mixed beech and oak forest of Villiers Cotterets, in France, at Nancy, and in the spruce and silver fir forests of the western Cotterets." There they "derived great benefit from what they learned, and their example has been followed by a number of forest officers from different provinces of India."[11] For more information on the subject Hough recommended an article on forest conservancy in India by Hugh Cleghorn.[12]

Hough served as Chief Forest Agent of the United States until 1883. He had predicted ten years earlier that "those who take an active interest in it [state forestry] now . . . will deserve and hereafter secure an honorable place in the annals of forestry." His goal to initiate a forestry program in the United States such as the British maintained in India did not see fulfillment in his lifetime but was taken up, also on the empire forestry paradigm, by Charles Sargent and Gifford Pinchot, among others.[13]

Charles Sargent and empire forestry

His successor, Charles Sprague Sargent, was born in Boston and attended Harvard University. He became Director of the Harvard Botanical Garden in 1872. He added the duties of Professor of Horticulture and Arboriculture and became Director of the Arnold Arboretum in 1873. His writing affected public perception of forestry and legislation on many levels. As a special exploratory agent for the 1880 census, he published the *Forests of North America*. This massive scientific survey of American forests alerted the public to the devastation caused by unregulated timber extraction and forest fires. Now largely forgotten, he also published a popular magazine entitled *Garden and Forest* in the critical years between 1888 and 1898. In this journal he edited weekly reports, new studies of trees and garden flora, reviews of new books related to the environment, and editorials concerning the progress of forest legislation. Empire forestry received particular prominence. *Garden and Forest* became a clearing house of ideas for legislative action. Throughout

---

[10] ibid. [11] Dietrich Brandis, as quoted ibid., 47.
[12] ibid., 46. For the article Hough recommended, see Hugh Cleghorn, *British Association* (1868): 90.
[13] As quoted in Steen, *History*, 20.

1888–1898 Sargent discussed, proposed, and promoted legislation that inaugurated the nation's first forest reserves.[14]

Against this background, Sargent's *Garden and Forest* served as a meeting of minds in the conservation movement in the late 1880s and 1890s. Each issue reviewed books of interest and carried columns and editorials on forestry. Letters from readers also reported atrocities across the American landscape, while Sargent alerted conservationists and congressmen to late-breaking preservation policies in other nations. Decades later Bernhard Fernow reminded readers that the journal *Garden and Forest* "should not be forgotten" for the role it played between 1888 and 1898 to "enlighten the public on forestry matters."[15]

International in scope, the discussion of forest literature in *Garden and Forest* offered panoramic visions of deserts reforested, wastelands reclaimed, and rain cycles restored. Sargent urged empire forestry as a paradigm on Congress, as it considered legislation to set aside forest reserves: "India has given to the world the most conscious example of a national forest policy adopted over a vast area . . . We can pick out climatic parallels between portions of India and of the United States more readily than we can between the United States and Europe."[16]

Why? Because India boasted evergreen forests like the northern territories of the United States, a great interior plain like the Midwest, and tropical areas like the Gulf Coast states. Moreover, the rail network in India crisscrossed the large subcontinent coast to coast. Even though forests were ravaged by speculators and contractors, the public in both India and the United States expected use of the forest by right. But Dietrich Brandis and Berthold Ribbentrop had "walked India" through the "education stage" and "recognized at once that conservative management could only be initiated by the government, the greatest landlord of the Empire."[17]

This aspect of the Indian example was critical to Sargent. Though he advocated government intervention, he also tried to promote empire forestry-style working plans adapted to the American market.[18] Because the public demanded forest access, Brandis and Ribbentrop made "settlement" with the public a priority, sorting out the thorny issues of what rights would be given out to whom, while keeping the ecological and economic value of the forests intact for the state. To gain legislative sanction and enforcement, they rejected the romantic notion of "a more complete ownership." Sargent expressed breathless admiration for the achievement of Brandis and Ribbentrop. By 1896, 130,000 square miles of Indian forests had "been formed into permanent forest reserves, in which the rights of the state and the adverse rights of the communities and private persons have been finally determined." The reserves dwarfed the forests of western Europe, with some working

[14] Biographical information on Sargent can be found in the *Journal of Arnold Arboretum* (April 1927) and the *National Academy of Science: Biographical Memoirs* 12 (Newark, 1929).
[15] Fernow, *Brief History of Forestry*, 371.
[16] Charles Sargent, *Garden and Forest* 9 (13 May 1896): 191–192.   [17] ibid.
[18] "An American Working Plan," *Indian Forester* 20 (1894): 238, 239.

plans for areas larger than Switzerland itself. Great Britain administered in her colonies a forestry area ten times her own geographic size, and it appeared to many Americans, Sargent among them, that empire forestry proved worthy of emulation.

Sargent recognized early on that the United States needed instructions in forest management and in models of legislation.[19] Intermixed with lamentations and alarm over the white pine forests in the south, the Adirondack forests in New York, the Douglas fir forests of the northwest, the pine forests of New Jersey, and the redwood forests of California, Sargent outlined in 1889 some basic points for American replication of empire forestry. "To provide for the conservation of the forest" Congress should withdraw all the forest lands belonging to the nation from public sale. The United States Army should be deployed for the "care and guardianship of the forests belonging to the nation . . . [Because] the forests are pillaged by settlers, and by the employees of railroad and mining companies, without scruple or limit," hence private exploitation necessitated constant and effective policing. Finally, the President should appoint a commission to examine the condition of the forests belonging to the nation's care and create a "comprehensive plan for the preservation and management of the public forests . . . a system for the training by the government of a sufficient number of foresters for the forest service" including a national school of forestry modeled along the lines of "the national military academy at West Point."[20]

There is reason to believe Sargent had India in mind. In 1887 he reviewed a government report by Ribbentrop, Inspector General of the Indian Forest Department.[21] After recounting the accomplishment of the forest department under Brandis and then Ribbentrop, Sargent suggests that

The history of forest administration in India might be studied with advantage by the Secretary of the Interior and members of Congress of the United States. [Unlike India] the forests which grow upon our national domain produce no income. The land upon which they stand is sold sometimes at a mere nominal price, and while the government is waiting for customers the forests themselves are robbed of their best timber, burned, pastured, devastated, and destroyed.[22]

In addition, the Indian example showed Sargent how to build a management system from scratch, drawing on medical and military personnel to "bring such knowledge to bear on the question." The Ribbentrop report gave Sargent hope that American schools of forestry, yet to be founded, could attain world preeminence, like Cooper's Hill in England, or even the forest school at Dehra Dun founded in

---

[19] Charles Sargent, "The Future of our Forests," *Garden and Forest* 1 (1888): 25.

[20] Charles Sargent, "The Nation's Forests," *Garden and Forest* 2 (1889): 49.

[21] B. Ribbentrop, *Review of Forest Administration in British India for the Year 1885–86* (Dehra Dun, 1887). Sargent's review included a discussion of Lt. Col. Ian Campbell-Walker's *Report of the Forest Department, Madras Presidency, for the Year 1885–86* (Dehra Dun, 1887).

[22] Charles Sargent, "Periodical Literature," *Garden and Forest* 1 (1888): 48.

1878. Though far from Europe, the Dehra Dun school, he observed, "ranks with the best institutions for education and subordinate staff in any country in the world."

But revenue impressed Sargent the most. He noted that in India revenue

has more than kept pace with the growth of the expenditure. The net average surplus for five successive five-year periods beginning with 1867–68 and ending with 1891–92 is as follows: (1) 1,339,000 rupees: (2) 2,129,000 rupees, (3) 2,689,00 rupees, (4) 3,848,000 rupees (5) 6,186,000 rupees with the cash surplus for the year 1881–82 being about seven and a half million rupees.[23]

Cash surplus not only augmented the government budget in India, it superimposed huge industrial and economic activity that public use of forest product engendered. In addition, the forest department even produced surplus revenue while defraying the expenses of the forestry schools, countless fire lines, forest houses for rangers, enforcement patrols, replanting, roads, canals, and railways to make the timber accessible. Development, conservation, and profit were concomitant with empire forestry. Sargent enthused that "we are certainly justified in taking heart and hope at what has been accomplished in India during thirty years," for here, as in India, the government holds large forest-bearing areas, and therefore "there is no reason to doubt the same thing can be done in America."[24]

The money saved by forest management in the US would be immense, Sargent believed, if Congress installed fire protection. For this he also turned to India and, nearer to home, Canada. In "Forest Fires – Another Lesson from India," Sargent adds that "The [Indian] Forest Department has thus proved clearly that it is possible to protect large forest areas from fire even in the very driest climate by a well-considered system of patrol."[25] Citing the *Report of the Commissioner of Crown Lands for Ontario*, he noted that Canada had already organized a system of fire protection that saved large stands of forest.[26] The cost of such fire protection came to three and a half cents per thousand feet of wood, which meant that if successfully emulated by the United States, fire protection for the states of Michigan, Wisconsin, and Minnesota would cost no more than $35,000 a year! Given that millions of dollars worth of timber burned to the ground every year, this fire protection would result in a massive saving for both industry and the government.

Sargent recounted how the president of the National Academy of Science had given three questions to the Secretary of the Interior that he felt America needed to answer regarding forestry policy. First, what proportion of the forests in the public domain ought to be privatized? Second, how should the government forests be administered? And third, what provision should be made for "a continuous, intelligent and honest management of the forests that have already been made"? Sargent himself proposed that "most of these questions have received an

---

[23] ibid.   [24] ibid.   [25] Sargent, *Garden and Forest* 9 (27 May 1896): 211.

[26] ibid. In this same article Sargent refers to the *Report of the Commissioner of Crown Lands for Ontario, 1895* (Ontario, 1896).

actual and practical answer in the management of the Indian forests." Public opinion can be educated to adopt and enforce legislation that will "look toward the selection of suitable tracts for conservative forest treatment" with India in mind.[27]

His ardent editorials had the desired effect, for to Sargent's delight the National Academy of Science asked Dietrich Brandis, former Inspector General of Forests in India, to lay out a plan of action for the protection of American forests that Congress ought to consider. Brandis responded by suggesting: (1) the collection of data, as had been done in India under the Survey; (2) the instigation of efficient timber extraction; (3) the initiation of discriminatory logging; (4) the planting of trees for valuable wood; (5) the replanting of trees for sustained yield; and (6) the reservation of as many large areas as possible into government forests. Sargent recommended these points to his readers and elsewhere concluded that "it ought not to be impracticable to frame a system of forest management for this country which would contain all the essential features of the plan which has proved such a conspicuous success in India."[28]

It is not the romanticism of the Lake Poets that Sargent and other Americans picked up from empire forestry, but rather a hard-headed "demonstrated use" argument for forest management that produced clear results for a variety of constituents, as well as a positive revenue for the government. In a quip that reverberated through the American press, Gladstone announced that the greatest obstacle to a sound forest policy in Great Britain itself, "was the superstition that invested trees with a certain sacredness so that felling was looked upon as a sacrilege."[29] The prime minister's observation, noted approvingly by Sargent, summed up the practical, utilitarian approach to nature that proved so successful for the British colonies. As Sargent pointed out, environmental progress is retarded when

Worthy people who, in their newborn zeal, are led to speak of all lumbermen as enemies of the human race. Of course there can be no system of forestry without tree-cutting, and the protest, to have any value, should be made against wasteful cutting or the stripping of mountains where the trees serve a higher purpose as a protection to the water courses than they can when made into lumber. It often happens too, that to secure the highest landscape beauty, trees . . . need to be removed.[30]

Gifford Pinchot and "what saved the national forests"

Gifford Pinchot, the third of the American forestry triumvirate, pioneered early American forestry and conservation. Born in 1865, he graduated from Yale and studied European methods of forestry at the National Forestry School in Nancy, France, and later in Austria, Switzerland, and Germany. After a stint as a private forester on the Vanderbilt estate, he worked with the National Forest Commission

[27] Sargent, *Garden and Forest* 9 (5 Aug. 1896): 191, 192.   [28] ibid., 312.
[29] Charles Sargent, "Arbor Day," *Garden and Forest* 1 (1888): 73.   [30] ibid.

of the National Academy of Sciences to craft a strategy to reserve large tracts of government land under the management of state forestry. In 1897 he served as Confidential Forest Agent for the Secretary of the Interior and in 1898 assumed the post of Chief of the Forestry Division, then, in 1905, he became chief of an independent forest service under the umbrella of the Department of Agriculture. He served with distinction under four presidents, including William McKinley, Theodore Roosevelt, and William Howard Taft, until his retirement in 1910. His other accomplishments include serving as member of the Public Lands Commission, the Inland Waterways Commission, and in 1908, Chairman of the National Conservation Commission. He founded and then became a professor of the Yale School of Forestry. His articles in popular magazines and the sheer volume of press coverage that his activities garnered gave him ample opportunity to explain to a reluctant public the need for governmental control of public forest lands.[31]

When Gifford Pinchot decided to be a forester, he went to Europe to study. What tends to be forgotten is that he studied primarily under French- and German-*trained* foresters, who in turn had served much of their professional careers in British India. Though unsure of the quiddity of his new vocation, he heard "that Forestry was practiced in British India, and it occurred to me that I might get some publications on the subject if I went to India House in London and asked for them."[32] This plan suggested itself to Pinchot because no journals of forestry had been printed in English other than the *Indian Forester*. Later, when American journals of forestry appeared, they imitated the *Indian Forester*, which remained the premier forestry journal until well into the twentieth century.

Accordingly Pinchot went to Wilhelm Schlich, at that time head of the forest school at Cooper's Hill in England, where young men were trained for the Indian forest service. The former Indian administrator promptly advised young Pinchot "to strike for the creation of National Forests" in the United States.[33] With a copy of volume one of Schlich's *Manual of Forestry* under his arm, he traveled to Nancy, where "for many years the foresters for the British Indian service had been trained." Afterwards he went to Germany to study with Brandis, to whom, he later claimed, "[I] owe[d] more than I can ever tell . . . After I came home I sent him news and many questions about what was doing and needed to be done in American Forestry . . . we never lost touch."[34]

---

[31] The outline of Pinchot's career can be found in his eminently readable autobiography, *Breaking New Ground* (Washington, DC, 1947). See also Harold T. Pinkett, *Gifford Pinchot: Private and Public Forester* (Urbana, 1970).

[32] Pinchot, *Breaking New Ground*, 6.

[33] Before he gained employment as a government forester, Pinchot implemented a "little working plan" on the Biltmore Estate in North Carolina, which he hoped would serve as a model of market-based forestry. See "American Working Plan," 239, 240.

[34] Pinchot, *Breaking New Ground*, 17. Gifford Pinchot, "Forestry Abroad and at Home," *National Geographic Magazine* (March 16, 1905): 375–388.

Brandis' hopes for Gifford Pinchot and for American forestry were well placed. In 1898 Pinchot assumed duty as head of the Division of Forestry and oversaw the transfer in 1905 of 63 million acres of forest land from the unmanaged public domain to the Department of Agriculture. His staff of 11 employees in 1898 grew to a professional corps of 821 in 1905. Under his administration the formation of the modern forest service took place.[35]

Few today realize how fascinating the public found forestry in the late nineteenth century. Arcane reports from Indian foresters received widespread treatment in American magazines.[36] If forestry had been a discussion solely about timber supply and the timber industry, it is doubtful that interest in the subject would have been so far-reaching. Rather, forestry was the flagship of early environmentalism and a fledgling ecology. Pinchot shared Ribbentrop's view of the forest as a "household of nature" and described it as a complex organism with "a population of animals and plants peculiar to itself." Additionally he saw the forest as "beautiful as it is useful."[37] Forest officials of the Indian Forest Department were even interviewed to understand the innovations. A prime source of information proved to be the *National Geographic*, which ran regular articles on forestry from around the world.

In an article in the *National Geographic* entitled "Forestry Abroad and at Home," Pinchot, then Chief of the Bureau of Forestry, stated that America had profited by the forestry so advanced in British colonies at large. "In Australia and New Zealand forestry has already made important advances. In Canada the English have made real progress in forestry." While Canada had retained full possession of the forests, it nonetheless sold off the surplus timber, guaranteeing a solid return from the land – land that it guarded with an efficient fire protection service. Hough had also admired British forestry in Canada and corresponded with Dietrich Brandis on the subject. He proposed a timber lease system comparable to that used in Canada, to be administered by the General Land Office.[38] From the Cape of Good Hope,

---

35 William G. Robbins, *American Forestry: a History of National, State, and Private Cooperation* (Lincoln, NB and London, 1985), 12.

36 Stories and interviews that appeared in popular magazines were often reviewed in forestry publications. For example see Sargent, "Periodical Literature," for a review of an interview Mr. George Cadell, former Indian forester, gave to *McMillan's Magazine* (January 1888).

37 See Gifford Pinchot, "A Primer of Forestry, Part 1, The Forest," *Direction of Forestry*, Bulletin 24 (United States Department of Agriculture, 1903). To illustrate how reviewers and the general public interpreted his work as environmentalism, see W. R. Fisher, "An American Primer of Forestry," *Journal of the Society of Arts* 54 (1905–1906): 730.

38 See Franklin Hough, *Report Upon Forestry*, vol. III (Washington, DC, 1882), 1, 6, 8, 14. Bernhard Fernow, Chief of the Division of Forestry between 1886 and 1898, helped draft the forest reserve legislation of 1891 and proposed "the Canadian plan" to protect forest fires. This is widely regarded by scholars to be the prototype of the fire protection programs administered under the Weeks Law of 1911 and the Clarke–McNary Act of 1924. Other works circulated that pointed Americans to empire forestry in Canada: the Department of Agriculture, Statistical Office, published a *Report on the Forest Wealth of Canada* (Ottawa, 1895), which served as an appendix to the *Report of the Minister of Agriculture for 1894* (Ottawa, 1895), circulated at forestry associations; J. C. Chapais' *Canadian Forests:*

"where they have an excellent forest service," to British India, where "they have met and answered many questions which still confront the American Forester," the British Empire had in thirty years created "a forest service of great merit and high achievement."[39] Pinchot ticked off British colonial forestry credits one by one and concluded that in comparison, "The US has scarcely yet begun."[40]

The concrete examples accomplished in the British Empire impressed Pinchot, Hough, and Sargent. All three Both men discerned that the empire forestry matrix produced a net revenue from the start. In a paper to the American Economic Association entitled "Government Forestry Abroad," Pinchot pointed out that the forests in India satisfied all the needs of the population without deforestation, and protected the water supply in the mountains as well. Additionally the Indian prototype served to show the United States how to proceed with government forestry in a country where "interference by the government with private rights would be so vigorously resented and where private enterprise must consequently play so conscious a part."[41]

The same opposition had existed in India, but the practical settlements under Brandis and Ribbentrop enabled forestry to progress. European land, monopolized as in Great Britain, or tenured under autocracies, as in the Russian Empire, could not present the same analogy. Neither could France, though advanced in forestry methods, because the government did not possess comparably vast areas of public land. Thus India had special significance for the United States as "the closest analogy to our own conditions in the magnitude of the area to be treated, [the] difficulties presented by the character of the country . . . the prevalence of fire, and the nature of the opposition which it encountered, [all these examples are] to be found in the forest administration of India."[42]

Pinchot believed that the precedents of empire forestry had saved the national forests of the United States. He stated in his autobiography that

Admirable as German Forestry certainly was, there was about it too much artificial finish, too much striving for detailed perfection . . . Dr. Brandis never let his pupils forget . . . that in the long run Forestry cannot succeed unless the people who live in and near the forests are for it and not against it. That was the keynote of his work in India. And when the pinch came, the application of that same truth was what saved the National Forests in America.[43]

Conclusion

Empire forestry posed for the United States both the environmental problems and solutions in stark relief to foresters, congressmen, and the public. Given the devotion to the free market in the United States, it is surprising that it followed

Illustrated Guide (Montreal, 1885) sold tolerably well in the United States. A. T. Drummond's *Forest Preservation in Canada* (Montreal, 1885), printed as an addendum to the *Report of the Annual Meeting at Boston of the America Forestry Congress* (1894) explored the discrepancy between empire forestry in Canada and the dearth of sound forestry policies in the United States.

[39] Pinchot, "Forestry Abroad and at Home," 375–388.   [40] ibid., 376.
[41] Gifford Pinchot, "Government Forestry Abroad," *American Economic Association* 6 (1891): 50.
[42] ibid., 50.   [43] Pinchot, *Breaking New Ground*, 17, 18.

22 Lumber mill in British Columbia.

after colonial countries in the reservation of forest areas. In Australia, for example, Lieutenant-General Ralph Darling established land-purchase rules that steered clear of broad farm ownership to favor grants of one square mile or more for "respectable" people willing to invest 500 pounds or more. Larger grants of up to 9,900 acres were sold at competitive bids. Conquest rather than purchase proved the rule in Cape Colony, assigned to Britain by 1815. Though European settlers could purchase land by sweat equity, the military nature of land acquisition beyond the Orange did not raise expectations of unlimited free or cheap land from the government. In Canada, due to the misguided effort to eliminate speculators, most new forest and agricultural lands were highly concentrated in few hands. H. G. Ward complained in the House of Commons in 1839 that land transfers were made through "personal edicts of the Secretary of State instead of under statute." By contrast American settlers expected land that could be purchased at Congress price, that is, $1.25 per acre, or if more, after improvements by speculators, which usually included a road, store, and often a bank and a church. Environmentalism in nineteenth-century America cut against the egalitarian and progressive grain, while the oligarchies established under direct and indirect imperial rule proved most compatible with the "settlement" of rights, precisely because fewer rights were distributed to fewer individuals.[44]

---

[44] For Australian land policy in the 1820s, see C. N. H. Clark, *History of Australia*, vol. II (Cambridge, 1962–1968), 69; For South African land policies in this period, see G. M. Theal, *History of South Africa, 1798–1828* (London, 1903), 202ff.; for H. G. Ward's speech to Parliament see *Hansard*,

23 Douglas fir and western hemlock forest in British Columbia.

Hard-headed environmentalists like Hough, Sargent, and Pinchot found a ready-made model to persuade the public and Congress that the reservation of vast areas of the public domain would simultaneously serve environmental, industrial, settlement, and budgetary purposes. The empire forestry matrix of government reservations, fire protection, professional management, and revenue-enhancing forests provided the solution to the tension between Romantic preservationists' notions

3rd series, 33, 852. To compare American intolerance of governmental interference with colonial tolerance, see Forester, "Conservancy of Forests," *Cape Monthly Magazine* 16 (1897): 166.

and *laissez-faire* policies. Fernow, chief of the Division of Forestry between 1886 and 1898, alludes to this when discussing the influence of Canada in the United States. He argued that Canada, "having escaped the period of sentimentalism which in the United States retarded the movement so long, could at once accentuate the economic point of view and bring the lumbermen into sympathy with their effort."[45]

This differentiates the men who implemented the forest reserves from men like Henry David Thoreau, John Audubon, and John Muir, whom environmental historians today celebrate as the founders of modern environmental thought. The careers of Franklin Hough, Charles Sargent, and Gifford Pinchot illustrate how the legislative process that resulted in the reservation of millions of acres required a dedicated effort of selling the proposal to a suspicious public little given to poetic raptures, especially in the western states.[46]

India exemplified how forests could be utilized by the public and not "locked up" like Yellowstone Park. Settlers and lumber companies could extract forest product while the state retained ownership. Thus western senators and congressmen acquiesced to the idea that the reservation of large forest areas would be in the best interest of the public. Empire Forestry denoted a less pristine solution that proved to be the practical compromise that both the early environmentalists and the public found acceptable. Empire Forestry, as the life and work of Hough, Sargent and Pinchot transcribe, laid the cornerstone of modern American environmentalism.

[45] Fernow, *History of Forestry*, 371.

[46] Pinchot's successor, Henry Graves, also "passed through the classic curriculum of British colonial forestry." Brandis considered Graves as well as Pinchot his protégé. See Pyne, *Burning Bush*, 262, 263.

7

From empire forestry to Commonwealth forestry

As the empire transformed into a commonwealth, so empire forestry transformed into Commonwealth forestry. Forestry and the institutions that had been built to protect the environment were handed intact to the westernized elites who controlled most of the newly independent states following the Second World War. The conservation movement broadened in the postwar years to include sweeping protections for wildlife, flora, wetlands, and issues of pollution and population control. The conservation movement, with empire forestry at its core, merged with other streams of thought and activities to form the core of a modern environmental movement that includes forestry as a central concern.

Empire forestry, transformed into Commonwealth forestry, is now lost in the broader environmental movement that it helped form precisely because of its success. Its conceptions of government ownership of nature, the right to settle land claims and reservation, the multiple-use doctrine based on Benthamite ideals of the greatest good to the greatest number, these conceptions are now so integral to environmental thinking that empire forestry has, by its success, been largely forgotten. As the empire changed into a commonwealth the massive inheritance from empire forestry has been harnessed in some cases, and improved upon and squandered in others. Once again the question of the rule of law has surfaced. Where countries that attained independence kept the law over the government official, and administered it responsibly, the environment has to that same extent been protected. Where the rule of law has not been maintained, this has not been the case, and the environment has suffered.

After the First World War the British government sponsored a forestry commission that uncovered Britain's desperate need for raw materials. During the course of the war Britain used up almost all available mature timber, and, if it were to provide the promised housing to returning soldiers and have the materials needed for economic growth, it need to find a way to meet the demand for timber. The Forestry Commission toured the British Empire and sponsored in turn the British Empire Forestry Conference. This conference took stock of the timber resources of the empire, then almost a third of the land surface of the world, and attempted to coordinate an empire-wide forestry policy. This conference has met intermittently and evolved from the British Empire Forestry Conference in the 1920s into the Commonwealth Forestry Conference after the Second World War. These conferences point to a consolidation and institutionalization of empire forestry.

The perception that Indian forestry practice deserved credit for empire forestry throughout the colonies was enunciated in the 1880s and 1890s, and then again at the first British Empire Forestry Conference held in London in 1920. In fact, this conference – which called together forestry and government officials from the entire empire – opened the session with the Lord Mayor of London explaining that "It has been found necessary to ask the forest services in India" to "come here and consult and assist us." By 1920 Indian expertise had such a reputation that an empire-wide conference to share information and ideas seemed germane, particularly after the shortages experienced during the Great War, which created a voracious market for wood products.[1]

But the Indian example would not be enough to establish a sound empire forestry policy. The report asked how much would have been saved "if 20 years ago we had had in Britain all the available information about trees of the Pacific Slope, Douglas fir, Sitka spruce, Western larch, etc.?" South Africa would have known about Australian gums, Canada and Australia could have taught much on logging organization, and Britain, with access to schools and laboratories on the Continent, could have shared even more knowledge. The 1920 conference estimated that Indian forest practice now managed 50 percent of the forest wealth of the world.[2] The Imperial Forest Service received students trained in forestry schools at Dehra Dun in India, Cooper's Hill, Kepong in Malaya, Ibadan in Nigeria, Arusha in Tanganyika, Kiterera in Uganda, Adelaide and Sydney University in Australia, and Toakai in Cape Town. In Britain, forestry schools were established at Oxford, Cambridge, Edinburgh, Aberdeen, and Bangor. (Empire foresters were either instructors or advisors for instructors at the forestry schools at Yale and Cornell.) These students, destined for every continent, transplanted Indian expertise to the world.[3] The conference helped the empire to unite its forestry program and essentially move from Indian forestry to empire forestry.

The conference confirmed the belief that Dalhousie's Forest Charter provided the framework that spread "during the last 60 or 70 years . . . [to] many of the other parts of the Empire . . . to the Federated Malay States and to New Zealand many years ago, even to Australia, to South Africa, to the West Indies, to the Sudan, to Cyprus, in fact, to any part of the British empire." That did not mean that each country would not find separate solutions to the political opposition aroused by

---

[1] *British Empire Forestry Conference*, 1–2.   [2] ibid., 14.
[3] ibid., 23; *The Training of Candidates and Probationers for Appointment as Forest Officers in the Government Service*. Report of a Committee appointed by the Secretary of State for the Colonies, July 1931, Colonial no. 61, 1931; Worthington, *Sciences in Africa*, 17.

new state forests, for the self-governing dominions especially did not have the power that imperial foresters enjoyed in India.[4]

The reports at the conference summed up the progress of each territory in establishing forestry policy and practice. For the Union of South Africa, C. E. Legat, Chief Conservator of Forests, reminded his audience that South Africa had the least amount of forests of any great Dominion in the empire, but had nonetheless inaugurated a vigorous forestry program.[5] Due to the paucity of existing forests, South Africa – unlike India or Canada – had placed greater emphasis on plantations and reclamation of forests – over 5 million trees were planted yearly for afforestation. A special forestry department operated under the Department of State, which in turn operated under the umbrella of the Minister of Agriculture, with headquarters at Pretoria and a training school at Toakai.[6]

Interestingly, the larger forests were often managed in smaller countries. M. Thompson, Director of Forests for Nigeria, and representing Nigeria, the Gold Coast, and Sierra Leone, reported that for Nigeria alone rain forest covered 10 percent of the country, with an additional 2 percent mangrove jungle, 8 percent monsoon forest, 45 percent savanna forest and only 30 percent desert, scrub, sand, water or cultivated land for agriculture. Here most of the forests were communally owned. The government spent much of its efforts working with tribal officials to practice forestry to ensure proper protection from fire and replanting. Even so the state managed to conserve 1.5 percent of the existing forests as absolute state property, while a training school for native staff in Zaria in the Northern Provinces (which was closed during the First World War) produced an indigenous core of foresters that were actively educating Nigerians about good forestry practice.[7]

Edward Battiscombe, Conservator of Forests for the British East African Protectorate, and speaking as well for Uganda, Tanganyika (formerly German East Africa), and Nyasaland, reported a "fairly comprehensive Forest Ordinance" with a "well-staffed forestry department whose policy was strict conservation and regulation of exploitation."[8]

G. E. S. Cubitt, Conservator of Forests for the Federated Malay States and Straits of Settlements, reported that they managed "a properly organized Forest Department" with reserves of evergreen forests that constituted 10 percent of the total area of the country, "and [that] we do not intend to stop anything short of 25 percent." As with forestry throughout the empire, "The organization of the Forest Department also follows closely Indian lines – no doubt owing to the fact that 20 years ago the Forest Department of the Federated Malay States was organized by an officer in the Indian Forest Department . . . The powers and duties of the various officers are similar to those in India."[9]

Representing British Guiana, the Bahamas, British Honduras, Trinidad, and other West Indian islands, C. S. Rogers, Conservator of Forests for Trinidad and

---

[4] *British Empire Forestry Conference*, 67.   [5] ibid., 41.   [6] ibid., 42.   [7] ibid., 47–48.
[8] ibid., 54–55.   [9] ibid., 56.

Tobago, discussed the implementation of forestry policy in these areas. British Guiana and British Honduras had yet to initiate forestry policy by 1920, but he recommended doing so with all speed. Trinidad and Tobago had 330 square miles set aside as forest reserves. In the Bahamas, with eighteen principal islands, the state owned 13 percent of the colony in forest land.[10]

The High Commissioner for New Zealand, Thomas Mackenzie, observed that New Zealand had set aside forest reserves in the 1880s, which now, after an intermediate period of regression, included 20 percent of New Zealand's forest area.[11] "There was a great deal of care exercised from the first in the preservation of... forests."[12] For Australia, H. R. Mackay reported that though Australia had few forests inside the coast lands, these had lately come under protection. He admitted that "In the early days of Australian administration and settlement, and indeed up to about 12 years ago, no State... developed... a forestry policy" of any consequence. But action had finally been taken to establish a national forest reservation that would reserve 1 percent of the continent of Australia, small in regard to the land area but a significant part of the forest area.[13]

E. H. Finlayson, head of the Forestry Branch of the Department of the Interior for Canada, reminded his audience that Canada was equal to the entire area of Europe, larger than the United States, thirty times the size of the United Kingdom, and involved 3,666 miles of travel by rail from the Atlantic to the Pacific. Unlike the large land give-aways in the United States, only 7 percent of Federal land had passed into private hands.[14] A Parks Branch operated parks within the Dominion Forests Reservation, with a forest service controlling more forest lands than any country in the world, including Russia, whose Siberian forests are not under management. Canada had 900,000 square miles of forests by 1920.

Many of the conservators at the conference expressed the belief that they were "in exactly the same state as India... when in 1855 Mr. Brandis was first sent out there, but we are now able to hold up the example of India to our Local Government."[15] Empire foresters also had great weight with the Colonial Office, as Mr. Ellis, the delegate from the Colonial Office, reminded them. He explained that "our general principle is to send out the best man we can find as Governor, and largely trust him." But lately, "in the last 25 years there has been a great change all over the world in the functions of Government. Twenty-five years ago you sent out a sensible man, and he worked by rule of thumb, like our Governors in the days of Lord Palmerston... but various branches of science now have a great deal to say to Government, and you are confronted on all sides with experts in medicine, in agriculture, in forestry, and a number of sciences." Along with the other established professions, forestry could by no means be ignored.[16]

The forest school at Oxford played a major role in Commonwealth forestry, as a training and research hub to this global wheel. The Commonwealth Forestry

---

[10] ibid., 60.   [11] ibid., 35–36.   [12] ibid., 36–37.   [13] ibid., 39.   [14] ibid., 26.
[15] ibid., 69.   [16] ibid., 71.

Institute began in 1905 as a school of forestry when Professor Wilhelm Schlich arrived to teach probationers (students) and prepare them for service in India. At first the India Office paid the salaries of the staff, and other expenses of the school, but later other colonial governments also paid to secure graduates for service in their respective regions. After 1918 the degree in forestry required a four-year course, two years of science with two years of forestry. Four professors early defined the program, giants in the history of forestry – Wilhelm Schlich, R. S. Troup, Harry Champion, and M. V. Laurie.[17]

After 1924 the school became a department and divided into two sections: the University School of Forestry and the Imperial Forestry Institute. The first section continued to produce forestry degrees, while the second, using money from colonial governments and the Forestry Commission in Britain, carried out research, advisory projects, and the training of students who already possessed degrees. In 1970 the school offered an undergraduate degree in both agriculture and forest science, to follow up on the increasing demand for plantations.

The school also offered an M. Sc. in forestry. From the 1920s to the present the Commonwealth Forestry Institute has provided a central hub of research and leadership for the Commonwealth, much of it in tropical areas. Forest officers from around the world, many with degrees from schools founded by empire foresters, nevertheless have undertaken special summer courses at Oxford to advance their training. The library is the world's most complete forestry collection in the world.[18]

Because each state in the British Empire had different climates, and also different political cultures, empire forestry evolved different styles and practices in different states. The move towards independence only accentuated these differences. India added a startling array of wildlife and nature reserves to the massive land set aside under the British. Africa progressed in certain regions and fell back in others. Australia continued to evolve effectively in the management of forests on its eastern edge, while New Zealand built impressively on the reservations begun earlier in the century to protect soil. Canada, for instance, placed a heavier emphasis on timber harvesting, falling behind many of the protective measures introduced in the United States. Despite the different paths of evolution each country has taken, the legacy of empire forestry in the commonwealth nations and the United States have remained central to the environmental project.

India

While the empire forestry model tried to project a system that paid for itself and aided the humane development of society by providing employment, needed

---

[17] Unnamed author, "Manuscript," green box, uncatalogued, back room of the Plant Sciences Library, Oxford University. The manuscript appears to be written by a faculty member of the Commonwealth Forestry Institute, and intended for publication in the *Oxford Times*, 1978.

[18] ibid., 3, 4.

infrastructure, soil, water, and wildlife protection, it did not boast that conservation would substantially contribute to surplus profit. Revenue enhancement meant, for Brandis and Stebbing, a self-sustaining department with benefits for the state and society, not necessarily a steady flow of cash profit. But this ethic did not go uncriticized.

Occasionally envious eyes were cast on the forest wealth. A particularly fearsome threat came in 1923 with the publication of a report by the Indian Retrenchment Committee. After the relative impoverishment of the United Kingdom during the First World War, Parliament, in an attempt to pay for social expenses at home, began demanding reductions in the cost of empire.

The Indian Retrenchment Committee forwarded to the Government of India recommendations for cost-cutting. Ostensibly, a reorganization between the central government and the Provinces prompted the need to cut expenses. But a marked reassertion of *laissez-faire* principles dominated the report, pitting "business sense" against the vision of conservation as practiced by the forest department. The report noted that the Government of India had in its possession "very valuable forest property" that the forest department had not fully exploited. The forest department had not given the committee satisfaction regarding "the quantity or values of timber . . . or stock on hand," and the committee saw the need for a major overhaul.

The report praised the department for its "high reputation for the management and development of the forests on technical lines" but insisted that in the future forests be managed "on commercial lines" with radical changes in "methods of administration." Research must be sharply reduced, to be "left appropriately to private enterprise," and, most tellingly, the committee recommended that "control of the Forest Department be vested in a manager with commercial experience in the timber industry."[19]

Fortunately, the suggestions of the report that concerned the forest department were not adopted by the Government of India. Similar ideas were revived during and immediately after the Second World War. Gross overcutting occurred during the war, resulting in severe wood shortages, charcoal shortages, and soil erosion. Foresters responded by attempting to hire more foresters, and to recruit more foresters by lowering the standards of admission to the forestry school at Dehra Dun. Foresters also recommended more strict regulations for privately held forests and the restoration of the forests held by villages that had overcut and oversold timber.

Independent India struggled with the same complaints from villagers and forest users as did the British elite. In justifying its need to place national needs over the community claim to local forests, the government pointed back to the forest policy of 1894, in which is asserted that adequate forests are necessary (1) for the wellbeing of the country, (2) for preservation of climatic and "physical conditions

---

[19] East India (Retrenchment Committee), *Report of the Indian Retrenchment Committee 1922–23* (London, 1923), 234–236.

of the country" and then (3) for forest products. The new government argued extensively that the state must protect national needs over and beyond local demands.[20]

As one example, the government argued for the protection of forests in the foothills of the Himalayas.

The State forests on the Himalayas are to be managed not solely in the interest of the people inhabiting that region alone, but also in the larger interest of the people living in the plains below. The right of the local population must, therefore, be regulated and its privileges of use restricted, even if some hardship is caused thereby, in order to prevent denudation of the Himalayan slopes, with its accompaniment of devastating floods, recurrent erosion of the fertile top soil, and deposition of coarse detritus on the fertile submontane tracts.

This is necessary because of the "immeasurable misery and loss on the unsuspecting millions" that would occur if no sacrifice is made. Grazing must be restricted, shifting cultivation discouraged, and the policy of *laissez-faire* curtailed. Otherwise "the balance of Nature" would be upset. This policy, with some later attempts to balance local demands with environmental concerns, is a contemporary issue very much alive in India today, and not merely a legacy of colonialism.[21]

Overpopulated India still derives most of its heating fuel from wood. The great cities, Calcutta, Delhi, and Bombay, despite many modern advances and the growth of high-tech industries, still derive approximately 30 percent of their fuel needs from firewood. Existing forests under management are therefore becoming thinner, with many areas of the "forest" unable to support an unbroken canopy and daily subjected to the pressures of continual thinning. This process of thinning, pointed out by the prime minister of India, Shri Morarji Desai, in 1979, began in the Western Ghats and the Himalayas, and has spread to many forest areas of India.[22]

The public has heard of community forestry through the advocacy of the Chipko movement. The Chipko movement began in 1972 when villagers in the Chamoli district actively protested against the harvesting of willow trees by the Simon Company. Throwing their arms around the trees that the company had marked for felling, they dramatized local resistance and caught the imagination of many. Women at Joshimath and other protesters at Dehra Dun intervened in harvesting and broke up timber auctioning.

Scholars interpret colonial discourse differently to explain the rise of scientific forest management in India. Some like Richard Grove suggest that imperial officials appropriated existing royal practices from the Indian and Mughal rulers who protected teak trees for royal use. Others argue that when such scholars give Indian roots to forest enclosures they "orientalize" environmental abuse – masking the European roots of forest conservation as imperial state building. N. L. Peluso

[20] *Explanatory Memoranda for the Central Board of Forestry, May 1951*, 24.   [21] ibid., 26, 30.
[22] Krishna Murti Gupta and Desh Bandhu, eds., *Man and Forest (a New Dimension in the Himalaya): Proceedings of the Seminars Held in Shillong, Dehra Dun and New Delhi and Organized by Himalaya Seva Sangh, Rajghat, New Delhi* (New Delhi, 1979), ix.

recounts how Dutch foresters justified state control by reference to the forest protection policies of Javanese kings and sultans. This, she claims, is entirely invalid because the Javanese conception of property "differed vastly from the notion of exclusive state property."[23]

The Indian Forest Act of 1878 provided for a category called the village forests, which, after the settlement of rights, were established for the villagers. The provision still operates in India today and is the basis and precursor of the "Panchayat forests," commonly referred to as the "community forest." J. P. Joshi, Additional Chief Conservator of Forests in the United Provinces, argued at a forest conference in 1979 that the Panchayat Forest Rules, framed in 1931 under a Scheduled Districts Act of 1874, allowed local magistrates control of the forests. He states, "As is well known these rules were mostly utilized for the unsystematic exploitation of existing forest growth resulting in the ruin of most of the Panchayat Forests."

While some local control has proved workable, "unfortunately the examples are very few." The Panchayat approach "has thus failed." He concludes that "the fact cannot be denied" that "areas which had been declared as reserved forests in the hills before 1921 still continue to have forest cover, generally in a healthy state."[24]

The construction of a "community forestry" ideal in precolonial villages has been regaled as a western oversimplification and an orientalist perspective. Echoed in the writings of Henry Maine, Karl Marx, and Max Weber, scholars imagine a precapitalist age where forest villagers constituted a utopian "other," by which to compare and critique conservation efforts. Giving a "merry India" derivative for community forestry, it replaced the golden age of Mughal conservation with a golden age of village conservation.[25]

The debate over indigenous versus colonial discourse may mask the power relations – and thus the bias – of scholars themselves. Gadgil and Guha appear to condone "grave inequalities of caste and class . . . [in] pre-colonial Indian society" because these inequalities "had a considerable degree of coherence and stability."[26] J. C. Kala, on the other hand, an official with the Ministry of Environment and Forests, cautioned that the older social structure of Indian villages has "practically ceased to exist," and a return to precapitalist social arrangement would now be impossible.[27]

---

[23] See N. L. Peluso, "The History of State Forest Management in Java," *Forest and Conservation History* 35 (1991): 69.

[24] ibid., 94–95. The author did not argue that community forestry was a failure in all instances. He pointed to the success of community forestry in Switzerland, where locals did not use wood as the primary source of fuel.

[25] See S. Nigam, 1990, "The Making of a Colonial Stereotype – the Criminal Tribes and Castes of North India," part 1 of "Disciplining and Policing the 'Criminals by Birth.'" *Indian Economic and Social History Review* 27:2 (1990): 131–64.

[26] See M. Gadgil and Ramachandra Guha, *This Fissured Land: an Ecological History of India* (New Delhi, 1992), 114.

[27] He also argues that the massive population explosion results in such a low per capita return on forest use that community forestry is of negligible benefit to villagers. Thus community forestry results

Whatever the position taken on indigenous versus colonial discourse, the similarity of forestry techniques and exploitation in modern-day India must be stressed. Village communities engage in specialized tree plantings and species hierarchy nearly identical to British conservancy for crop management. Fines and punishments enforce community forestry, as with state conservancy.[28] And unfortunately, the Chipko movement (the tree-huggers) has, as K. Sivaramakishnan points out, turned into a subversion of the Forest Conservation Act of 1980. Many of the local authorities empowered by the Chipko movement now engage in forest clearance along the very lines of profit that it originally opposed.[29]

No ideal seems to escape the inevitable balance between harvesting trees and environmental protection, because all involve a type of "settlement" of local rights similar to the arrangement under the Indian Forest Service. As a recent study indicates, the "settlements" under the Indian Forest Service in 1886 did not so much proscribe local rights as formalize them, protecting the whole community from local elites.[30] Fikret Berkes and Kerril and Ian Davidson-Hunt caution that "investing power at the local level" allowed "some groups to dominate local institutions" unfairly and that government policy should be devised "to protect villagers from caste and upper-class abuse."[31]

Conversely, a survey by J. Mark Baker suggests that "slightly unequal patterns of wealth distribution" contributed to the success of a community forest. "The conclusion regarding the effects of wealth distribution on collective action runs counter to approaches which suggest that egalitarian communities are more likely to cooperate." The question of indigenous versus colonial forestry then takes on a new light, pitting profit-taking by local elites in village forestry against profit-taking by a state that is committed – at least nominally – to egalitarian and public concerns.[32]

Community forestry can devastate a forest area, as many Indian foresters after independence have pointed out. An expanding population, now soaring past 1 billion people in India and likely to overtake the population of China in a few years, means that hordes of new people move into areas previously uninhabited. When these newcomers demand local control, it creates not only a daunting legal question, but places unsustainable demand on the ecosystem. Many areas that the

in not preservation, but "equitable removal by all." See J. C. Kala, "People's Participation in Public Wastelands Development in India: an Analysis," *Indian Forester* 120 (1994): 555–558.

[28] See R. Guha, *The Unquiet Woods: Ecological Change and Peasant Resistance in the Himalaya* (New Delhi, 1989); G. B. Pant, *The Forest Problem in Kumaon* (Gyanodaya Prakashan, 1922).

[29] K. Sivaramakrishnan, "Colonialism and Forestry in India," *Comparative Studies in Society and History* 37: 1(1995): 32.

[30] Fikret Berkes, Ian Davidson-Hunt and Kerril Davidson-Hunt, "Diversity of Common Property Resource Use and Diversity of Social Interests in the Western Indian Himalayas," *Mountain Research and Development* 18 (1998): 19–33.

[31] ibid., 30–31.

[32] J. Mark Baker, "The Effect of Community Structure on Social Forestry Outcomes: Insights from Chota, Nagpur, India," *Mountain Research and Development* 18 (1998): 61.

Chipko movements managed to wrest from government control have been ruthlessly sold off and the profits used up without the benefit of replanting or further management.[33]

The Second World War served as a bridge that deconstructed the old empire forestry standards of management and led to the difficult years after independence. Soon after independence came a call by Indian foresters to commercialize the whole forest estate. This latter suggestion depicts the mood of the new governing class after independence, when the overcutting out-paced even the abuse of the forests during the war. Marxist and state building ideas overcame Romantic and ecological considerations.[34]

A conference of senior forest officers held at Dehra Dun in 1946 foreshadowed the future. One paper, read by V. S. Kuppuswamy, argued that "an immediate and detailed survey" should be undertaken to determine what the forest department could supply, at what quantities and cost, with a "central pooling of information" about exactly which industrial requirements the forest administration met. There should be "close collaboration" between the industries and supplies department, particularly in the formation of working plans, to remake them "purposeful from the point of view of . . . Industry."

Immediately after independence, overcutting again picked up and a free-for-all pillage of the state forests ensued, with incalculable loss to the forest estate. The First and Second World Wars both served as opportunities to massively overcut the very forests that had been so carefully preserved and managed. The democratic allies saw forests and forestry as a strategic weapon for waging war. For this the colonies, including India, paid a heavy price. But if the war proved to be a watershed that abandoned largely the rational principles of forestry and conservation, independence saw a fresh round of environmental abuse. The postindependence elites largely ignored a forest policy identified with the old empire. The new control over natural resources empowered the new elite when the machinery of government fell into their hands. Further, rising population, industrialization, nationalism, and policies of industrial development, fortified with western aid, led in some cases to a wholesale abandonment of environmental management in the forests.[35]

Official pronouncements on the loss tend towards the arcane and understated, masking the extent of the problem in the early fifties. M. D. Chaturvedi, Inspector General of Forests, complained in 1951 that while the old Indian Forest Service "maintained a tradition of hard work, professional competence, and unsullied intellectual honesty," the reverse had occurred in recent years due to "ill-formed

[33] ibid., 297, 299, 300.

[34] *Proceedings of the First Senior Forest Officers Conference held at Dehra Dun from the 5th to the 7th April 1945* (Simla, 1946), 90, 92, 113, 105; *Reconstruction Committee of Council Second Report on Reconstruction Planning* (Delhi, 1944), 43.

[35] Gupta and Bandhu, *Man and Forest*, ix; *Forestry in Britain: Reconstruction Committee's Report and its Applicability to New Zealand*, (Wellington, 1919), 16.

pressure from different directions." He then suggested, unhelpfully, that "it is impossible to conceive that the present popular leaders of Indian Union would permit any incompetence and negligence in future."[36]

Due to increased political pressure, states were given the right to enact their own forest policy under the National Forest Policy of 1952. Some states had the right to assign reserved forests to a lower level of protection – a village forest. But the widespread practice of dereservation, usually resulting in the diversion of forest land into farms, forced the government to pass the Forest (Conservation) Act of 1980, where only the central government has the power to authorize dereservation. This, coupled with a moratorium on the felling of trees over 1,000 meters and a new 1988 National Forest Policy Act, has helped set aside further forest areas, prevent erosion and denudation, and involve the local community in forestry aims. The goal, difficult to reach with mass overpopulation, is to place one-third of the Indian subcontinent under forest cover, with two-thirds of mountainous regions under forest cover, and to link forest and wildlife areas with "corridors" for genetic continuity.[37]

In 1977 Dalhousie's principles were enshrined in the very wording of India's constitution, which reads "The State shall endeavor to protect and improve the environment and to safeguard the forests and wildlife of the country."[38] The challenges, however, are many. Management since independence has been sometimes excellent, sometimes bad, often corrupt and destructive to the areas that the agency is designed to protect. Often local people are ignored and so popular support is lost. Today less than 1 percent of the country is primary forest. With a rising population, and with a large portion of the nation's energy needs still coming from firewood, the prospects for the survival of forest land in India are grim.

If forestry has suffered, real progress has been made in the protection of wildlife in India. The India Wild Birds and Animals Protection Act of 1912, and then, further, the Hailey National Park Act of 1936, began the process of extensive protection. Project Tiger is a well-known example of successful wildlife protection in independent India. Tremendous strides were made since independence, setting aside further wildlife areas for birds, tigers, and marine ecosystems.

Corruption, however, threatens to undo these gains and to undo the path-breaking gains of empire forestry in the nineteenth and early twentieth century. R. K. Roy asks if "we [have] even maintained pre-Independence forest levels?" and answers in the negative, largely because of the decay of forests that are under protection, with 22 percent officially forested but which he considers to be "only 11% to 12% ... forested."[39]

Unfortunately, great controversy surrounds the implementation of a workable solution. Some argue, along the lines of Dalhousie, for a strict return to

---

[36] *Explanatory Memoranda for the Central Board of Forestry, May 1951,* ii, iii.

[37] *Government of India (n.d.) Department of Environment: a Profile,* Government of India, New Delhi, 23.

[38] Article 48a.   [39] Gupta and Bandhu, *Man and Forest,* 136.

24 Close-up picture of giant snails on a pometia tree in New Britain. Photo by
H. G. Champion, 1957.

scientific forestry and wildlife management, and others, like the socialist R. K. Roy,
propose "farming wildlife." Roy states that "Properly organized, all our sport-
ing birds and animals can provide both finance and part-time activity for local
Panchayats." Nature, he argues, can be sold or leased to local villages, where
community forestry and wildlife management will bring profit to the locals and
cut out the part played by large corporations. The record, however, of socialist
environmentalism in the former Soviet Union gives rise to concerns regarding
this utilitarian Marxist approach.[40] India is the land that gave birth to environmentalism. Paradoxically it may also be a land that will ruin the impressive environmental structure that once served as the model of nature management for the
world.

[40] ibid., 138, 139.

Africa

In South Africa, from 1895 to 1965, new areas came under protection at a rate averaging one new addition per year. Approximately 5.8 percent of the land area of South Africa is managed by the state for conservation purposes. The country is divided into the provinces of the Cape, Transvaal, Natal, and the Orange Free State, each with a separate conservation agency. A National Parks Act and a National Parks Board oversee the separate provincial agencies. These umbrella organizations were established by Parliament in 1926. The Department of Forestry, today called the Directorate of Forestry, is under the Department of Environmental Affairs.

The Department of Forestry, created in 1910, oversees 115 separate forest areas. Always with limited forest in South Africa, the department protects mountain catchment areas as well as forests set aside for timber harvest, tourism, and the protection of historic scenic areas. Natural forests cover 1.9 percent of the land area of South Africa. Plantation forestry still provides a large share of the timber needs and protects natural forest areas from exploitation.

Since 1983 the state has amended the National Parks Act to create partnerships with local authorities and landowners. The act allows the state to establish biosphere reserves, upgrade lands administered by the Defense Force, and add further catchment and conservancies in the Transvaal and in Natal, as well as to protect the oceanic islands of the Prince Edward Islands. All departments involved in conservation now work with the Council for the Environment to educate the public about the protection of coastlands and wetlands, and about urban planning. A "National Environmental Policy and Strategy" aids in the cooperation and national planning of conservation.

In Nigeria, forests were in particular need of protection because of threat posed by the rise of the automobile in the West. Automobiles need tires, and tires are made of rubber. The west African forests were being "creamed" (thinned) for the best mahogany and tapped for latex – the best coming from *Funtumia elastica*, or Lagos Rubber. Previously an attempt at conservation had been made by a rule that required twenty trees planted for every one harvested, but foresters soon saw the futility of trying to preserve the forests without having clearly demarcated areas under close management.

Two fortunate occurrences helped save the forests of Nigeria from complete destruction. The para tree became the primary producer of latex, which meant the bottom fell out of the market for Lagos rubber. Then in 1908, the chief forester aided the passage of an ordinance based on the same legislation that had established forestry in upper Burma. The principles made it possible to set aside a permanent forest reserve. This legalization determined the pattern of Nigerian forest laws for many years to come, and by the First World War increased the square miles of forests under reservation from 37 in 1903 to 1,346 in 1915.

The catalyst for this change was H. N. Thompson, a graduate of Cooper's Hill. Thompson had twelve years' experience as a forester with the Indian Forest Service in Burma. The government appointed him Conservator of Forests in 1903 and paid him a large salary of £1,000 pounds a year, a very large sum, that paid for "a widely read, cultured and able man" who had won the trust of the governors serving in the colonies. Nigerian forestry was "founded upon his knowledge . . . and upon the policy he so clearly enunciated."[41]

In 1910 Thompson succeeded in separating the agricultural and forestry departments so that as chief conservator he could concentrate entirely on the development of forestry without answering to the needs of an unrelated bureaucracy.[42] In 1914 the regions of north and south Nigeria united and Thompson functioned as advisor to the governor-general, with the title Chief Conservator. Regional game laws were enacted for the eastern section in 1916, later applicable to western and northern Nigeria in 1928 and 1963.[43]

Forest reserves were expanded during the 1920s and 1930s, with the Fauna Preservation Society of London pressuring Nigeria to increase game reserves as well. Reserves rose slowly to 29 in number by 1960. These reserves owe their origin to the Nigerian Forest Department, which in 1946 designated certain forest tracts as unexploitable. As of 1992 these reserves were 4.5 percent of the total area of Nigeria.[44]

Keeping forests reserved for climatic conditions persist to the present day. In 1964 J. E. M. Horne reiterated the necessity of preserving forests for the sake of water as well as climate. Ten percent of Nigeria in the 1960s was under forest management, and the creation of forest reserves was justified "on the grounds of hypothetical forest influences rather than of potential productivity."[45]

In Ghana, four stages characterize environmental practice, building upon the foundation of empire forestry from 1874 to 1939. During the first phase of formal forestry, a string of forest reserves protected 20 percent of the forests of Ghana, and today this now makes up almost wholly the remaining forests. The second phase, "timberization," lasted from 1940 to 1953. This period saw two events that overruled the rational working of the forests under empire forestry and engendered overcutting: the Second World War and the movement for independence. After independence, local chiefs were given far more say in the harvesting of wood from forest reserves and "community forestry" policies led to massive overexploitation.[46]

The third phase, the "diktat phase," lasted from 1954 to the early 1990s. Government officials pushed timber harvests in postindependent Ghana as a prime mover for development. Harvests were localized by a plethora of local companies, who paid low royalties. Finally, the fourth stage, the "collaborative phase," began

[41] D. R. Rosevear, *Vegetation, Forestry and Wild Life in Nigeria* (1953), 175.   [42] ibid., 179.

[43] World Conservation Monitoring Centre, *1992 Protected Areas of the World: a Review of National Systems* (Cambridge, 1992), Nigeria.

[44] ibid.

[45] *Proceedings of the First Nigerian Forestry Conference: the Role of Forestry in the Economic Development of the Savanna Areas of Nigeria* (Kaduna, Northern Nigeria, February, 1964), 61.

[46] Nii Ashie Kotey et al., *Falling into place* (London and Legon, Ghana, 1998), 3.

in 1994, when donor development agencies like the World Bank, the UK's Overseas Development Fund (ODA; now DFID), and the Ghana government changed strategy to reduce the export of logs and the annual cut, to work toward a more permanent forest estate. This rethinking of strategy also includes greater concern for wildlife and a return to the early empire forestry policy of taking into account the whole household of nature.[47]

Small patches of forest outside the reserves remain, and these are often fetish groves, or sacred areas around temples or with special mythical importance. But sacred areas contribute little to forest conservation or broad environmental goals, not because they do not have a valid local effect, but because they are simply too small. A recent report on forestry in Ghana is not optimistic. The "wave of deforestation has only recently arrived," its author laments, and arrived *"en masse"* with maximum pressure to tear down the boundaries so carefully built up under colonialism. The forest is being "nibbled away" at the edges, with abscesses in the shape of plantations, mines, and farms destroying the integrity of the canopy. A few sacred groves, the report emphasizes, will not change that.[48]

The empire forestry reserves established have proved the only protection for the forests, before and after independence. One-third of Ghana's forest estate was lost in just seventeen years, between 1955 and 1972, and satellite photos today show that the 20 percent of the forests reserved under the colonial forest department are the forests left today. The boundary lines established by empire foresters run in straight lines between pillars, and make a graphically clear outline enclosing the forests. H. N. Thompson, who served in Ghana as chief forester, wrote in 1910 that few forests would remain if reserves were not implemented on the Indian model. His prediction proved remarkably true. From space, his handiwork is starkly visible today – 80 percent of the forests outside the reserves have been destroyed.[49]

Australia and New Zealand

In Australia the decentralization of power slowed but did not stop the development of sound conservation and environmental policies. Up until 1901, each colony had a separate colonial government with separate strategies. Then in 1901 the states formed a federation of independent states. In this federation the states retained broad power over land and resources, including forestry policies. Federal forestry in Australia exerted influence by hosting centralized meetings, sponsoring publications, and dispersing funds.[50]

The national goals in 1920 set for the reservation of forests were not entirely reached until 1965, due to the lack of funds and the problem of mapping. After the First World War large areas had to be demarcated, particularly in Tasmania. The

---

[47] Ibid., 3, 4.   [48] Ibid., 8, 11.   [49] Ibid., 5, 11.
[50] Carron, *History of Forestry in Australia*, xii.

forest service's location in the Lands Department also slowed the acquisition and management of reserved forests. While all the forests on crown lands were considered state forests, it took decades of struggle over funding, personnel, and conflicts with other agencies to meet the goal set in 1920 of responsible management.[51]

Recognizing the need for an Australia-wide approach to forestry issues, the parliament of the Commonwealth government passed authorization for the Forestry Bureau in 1930. This new federal bureau oversaw forestry in the Commonwealth, educated professional foresters, researched methods of forestry management, and managed funding between the Commonwealth and the states. In 1935 the Forestry (Amendment) Act placed limits on the planting of exotic species (such as coniferous trees) and established "national forests" on the Indian model of tiered management. The multiple use of forests for timber, watersheds, recreation wildlife and grazing were also implemented.[52]

The Depression made it possible to employ thousands in labor camps to clear trails, build fire lines, map, plant trees, and otherwise manage the forest estate. The Second World War made great demands on the forests, however. Postwar policies worsened the environmental dilemma by increasing the consumption of wood for a housing explosion and industrial expansion that counterbalanced the consumer deprivation of the war. The result: further pressure to harvest wood at levels that were more ecologically unsustainable.[53]

One unforeseen event hastened the protection of forests. In 1939 a severe forest fire, the worst in living memory, killed 71 people and drew attention to the need for increased fire protection. Amendments to a 1916 act were passed and greater attention focused on fire-fighting techniques, many of them learned from both India and the United States. Partly as a result of better protection, revenues increased gradually throughout the 1950s and then shot upwards in 1965.[54]

Part of the multiuse conception of forestry included the preservation of forests for scenic beauty. Carron writes, pointing out the multiple-use origins of the national park:

As elsewhere in Australia in the late 1960s, the demand for both passive and active forest recreation began to increase [in Western Australia] . . . and, like forestry services elsewhere in Australia, the Forests Department moved to increase recreation as best it could . . . The inclusion of recreation as one of the multiple use objectives in 1976 provided an opportunity for . . . expansion.[55]

In Queensland the reservation of national parks began in 1908. Throughout the 1960s citizens expressed concern that a "national parks" policy be implemented, and this occurred in Tasmania with the National Parks and Wildlife Act 1970.[56]

The meaning of conservation evolved in Australia from the earlier emphasis on wise use and forest utilization to, in the 1970s and 1980s, a view that forests

---

[51] John Dargavel, *Fashioning Australia's Forests* (Melbourne, 1995).   [52] Ibid., 250.
[53] Ibid., 19.   [54] Ibid., 7, 70.   [55] Ibid., 175.   [56] Ibid., 92.

should not be "used" at all, but rather preserved for their own sake. Hal Wooten stated the change poetically when he explained that conservationists regard earth, with its

rich variety of plants and animals and the only known habitat for life in the universe, as a very precious place. We consider that our human position of intellectual and technological preeminence gives us not a right to destroy the planet and other forms of life on it but a responsibility to protect them. We have an obligation to act as trustees for future generations and for other species, and to endeavor to reconcile our interest with theirs in a reasonable way.[57]

When overseas markets developed for Australian wood chips, the Australian Conservation Foundation opposed using the state forests for further export. But the market for wood chips became irresistible. The ability to grind logs into chips and ship them cheaply to markets in Japan opened up new forest areas for exploitation. While public support grew for conservation on one hand, state governments raised the harvest quota on the other. Vast areas of forest land were cut in the 1980s and 1990s for wood chips.

Under the empire forestry model, harvesting of trees occurred under "Special License" that foresters carefully calculated to fit the working plan. This preserved a fair degree of ecological viability if the working plan were sound. In addition, since the empire forestry model was clearly tiered, the forester had less chance of being "captured" by local interests and gave permission only for the proper amount of harvest. But when the harvesting amounts could be set by separate state governments in Australia, free of a strong central authority, the result could be devastating – as seen with the decision to overharvest for wood chips.

In New Zealand most of the crown forests had been conserved, protected by a Forests Act of 1921–22. Forestry in New Zealand reflected the need to protect a supply of timber, water flow, wildlife, and provide for recreation. The same issues that motivated officials in New Zealand to protect the forests initially were still primary concerns in the 1930s and 1940s, with soil protection and flood control still a problem. Hawkes Bay suffered devastating floods in 1938 and Parliament appointed a committee in 1941 to investigate what further progress could be made in securing proper protection through forest cover.[58]

Legislators responded with the Soil Conservation and Rivers Control Act (1941). This legislation created a central council to oversee catchment areas and advise on policy and supplement needed work with funds. Engineering works occupied a great deal of the actions, and slowly the realization dawned that to safeguard these areas logging operations had to be severely curtailed and managed. The Forest Act of 1949 gave ministers the power to obtain and manage land for the purpose of conserving water and stabilizing the soil. The Soil Conservation and

---

[57] Hal Wooten, quoted ibid., 81.
[58] Peter McKelvey, *Steepland Forests: a Historical Perspective of Protection Forestry in New Zealand* (Canterbury, 1995), 165, 166.

River Control Amendment Act of 1959 further protected these catchments and addressed concerns.[59]

Since the Forests Act of 1921–22 patrols were introduced to protect against outbreaks of fire. Damages were slight until in 1937 a forest fire destroyed 40 million board feet of timber. Until this time the forest service could justly claim to have protected forests adequately from fire. In the 1950s better methods evolved to improve fire protection. Watchtowers were built over exotic forests, aerial patrols were set up with modern communication equipment and new firefighting vehicles equipped with pumps and mounted on four-wheel drive vehicles distributed.[60]

In addition to fire control, the forest service had to grapple with politically sensitive issues such as the control of wild animals. At first the introduction of exotic animals like pigs, beavers, opossum, and red deer met with the approval of many foresters. But it soon became apparent that the overpopulation of these animals was highly destructive to the soil. Red deer, for instance, would eat off the vegetation, including saplings, killing the regeneration of a forest. In catchment areas that needed the soil and water supplies protected for cities and towns, the issue became critical.

While the government assumed control of "noxious animals" in 1930, it was not until after the Second World War, in 1956, that control of these animals was handed to the forest service. An inventory of indigenous forests undertaken between 1945 and 1955 revealed a tremendous degree of damage from deer and pigs, among others, that had to be addressed. Accordingly the forest service licensed and oversaw the hunting of these animals. An export business for venison and wild pig, established to ship meat to West Germany and the United States, helped in the 1960s to maintain lower populations.[61]

The Forests Act of 1949 replaced the Forest Act of 1921–22 and concisely summed up Brandis' "multiuse" forestry concept eighty years later.[62] Multiple-use

---

[59] F. W. Foster, *Forestry in Relation to Soil and Water Conservation* (New Zealand Institute of Foresters, 1954), 1.

[60] F. Allsop, *The First Fifty Years of New Zealand's Forest Service* (Wellington, 1973), 76.

[61] Ibid., 84–87.

[62] It reads:

(a) All Sate forest land, whether for the production of timber or other forest produce, or for the protection of the land with a view to water conservation or soil stabilization, or for ensuring the balanced use of the land, or for scientific purposes, or for recreational or amenity purposes not prejudicial to forestry

(b) The establishment, culture, and maintenance of forests on State forest land, and the harvesting, utilization, transport, sale, or other disposal of forest produce from State forest land

(c) The granting of licenses, leases, permits, and other rights and authorities under this Act

(d) The enforcement of the conditions of licenses, leases, permits, and other rights and authorities granted under this Act or any enactment hereby repealed

(e) The collection and recovery of all purchase moneys, rents, fees, royalties, charges, and revenues of the Service

(f) Generally the exercise of all powers, authorities, and duties conferred or imposed on the Minister or the Forest Service by this Act.

forestry in New Zealand allowed the government to emphasize those practices that were most compatible with the needs of a mountainous region with special flood and soil concerns. The Tararua State Forest, later a state park, illustrated this flexibility. In 1960 the objectives of multiple-use management for this forest were stated, to "achieve the optimum land use for the area," to "protect and conserve the vegetative cover with a view to minimizing soil erosion, regulating runoff and conserving domestic water supplies," and to "develop recreational and amenity values."[63]

The existence of separate departments for Lands and Survey, Forest Service, Wildlife Service, and the Commission for the Environment, came to an end in 1987. The government replaced these separate departments by the Department of Conservation. This created a central organization to protect and preserve the natural environment and protect resources for the future. Conservation in New Zealand, after spinning off a variety of environmental responses, came under one head again in the 1980s.[64]

Canada and the United States

As a movement, forest conservation in Canada had its strongest support from government officials and the public in the years preceding the First World War. This support declined after 1915 and did not begin increasing again until the 1960s. Though the Indian model inspired a massive set-aside of land in both the United States and Canada, the attempt for a full Dominion forestry service, that is, a Canada-wide forestry service with a strong central structure, never materialized.

In fact, since the Constitutional Act of 1982 gave Canada the right to amend its own constitution, the power of the provinces over natural resources has radically increased. Building on the popular concept of local community control, the native peoples of Canada will have greater access to the forests and forest products, opening up forest resources to even greater risk of exploitation. The park systems that had been treated as inviolate are of particular concern. With 90 percent of the forests of Canada formerly under strict government control, this recent erosion of the power of the state over nature poses a grave risk.[65]

The mass media in the 1920s and the 1930s gave little notice to forestry in Canada. Not only did the First World War place huge demands on the forests, but the depression that followed the stock market crash in 1929 cut government spending on conservation and put a premium on economic interest. The Second World War followed with more overcutting of timber.

Because of the further abuse of the forests under wartime conditions, in 1949 the Canada Forestry Act was passed that established national forests, to complement the already existing provincial forests. Forest products laboratories and forest

---

[63] Foster, *Soil and Water Conservation*, 168.
[64] *World Conservation Monitoring Centre*, New Zealand.
[65] A. Hackman, C. Steard and G. Francis, *World Conservation Monitoring Centre* Canada, 1.

experimental areas were established. Most encouraging of all, the act gave the federal government power to advise and fund provincial and private forestry efforts with a view toward conservation.[66]

On the whole, however, the forestry question in Canada is not a story of success when compared to many of the other Commonwealth countries and the United States. Reformers in the United States built on the example and innovations of empire foresters to create an impressive and unique national environmental program. But in Canada politicians and the timber industry turned the "multiuse" forest into the uni-interest of timber extraction. The dream of ecologically minded reformers, like Henry Joly and Alexander Campbell, the last commissioner of Crown Lands in United Canada, has not been realized.[67]

With the increase of provincial and indigenous power over natural resources, federal-wide control and protection has moved Canada further away from the old empire forestry model of centralized control. The environmental movement split between many who came of age in the 1960s, who emphasized control over pollution and a more pristine vision of forest management, and those conservationists who believed in a balanced and reasonable extraction of timber. The issue now rests largely in the hands of local governments, whose vision of conservation varies.

In the United States, the environmental movement that empire forestry jump-started took off rapidly and developed into a record of protection and innovation equaled by no other part of the world. By the end of the 1920s the innovations, high professional standards, and expertise flowed outward from the United States and informed and enriched the forestry work of the empire and then the Commonwealth. The flow of ideas and information had by the 1930s come full circle. A web of interaction circled the globe with foresters reading, lecturing, and learning from each other.

Conclusion

When and where did environmentalism begin? Environmentalism, in the sense of practical action, began in 1855 in India, with the Forest Charter. This charter defined the right of the state over "nature," proscribed private interests, and initiated a new system of forest management. Dietrich Brandis, a trained botanist and empire booster, devised a "multiuse" model of forest management for most of British India. This model reconciled the needs of peasants, businessmen, and environmentally prone administrators; a tripartite alliance between political reality, revenue enhancement, and climate theory. The scattered Indian reserves initiated in the 1850s grew to 32,000 square miles by 1901 and encompassed tropical and deciduous forests, grasslands, and semiarid savanna. By 1939 empire foresters managed half the forest wealth of the world.

The origins of effective environmentalism, though not devoid of romantic notions, are found in a distinctly utilitarian concern for proper use of societal

[66] Gillis and Roach, *Lost Initiatives*, 248.   [67] ibid., 262.

resources. Romantic ideas about nature certainly have shaped much – though not all – of post-1945 western environmental sensibility. But serious environmental measures that actually took place in the nineteenth and early twentieth century occurred primarily as a result of imperial forestry officials who took a worldview toward global environmental problems. While romantic and health concerns dominated post-1945 environmental thought, it can be safely concluded that utilitarian conservation dominated pre-1945 environmental thought, and that this utilitarian environmentalism originated in an imperial context.

This new imperial management of nature entailed a drastic expansion of state powers and the evolution of a new breed of scientific expert: the forester. Only in the British Empire did a multiuse forestry program first solve the tensions between *laissez-faire* notions and the need for far-sighted state management of resources. When the United States and other countries decided to legislate a similar program, they turned understandably to empire forestry and empire foresters. Though environmental historiography owes its origins primarily to North American scholars much influenced by Romantic nature writing, environmentalism, in the sense of practical policies, owes its origin to empire forestry.

Environmental history is not literary history. Though Wordsworth, Thoreau, and Morris did generate lines of thinking that contributed to an ecological outlook, they did not initiate the protection of mangroves in the tropics, evergreen forests in Kashmir, Douglas fir and cedar in Canada, or pine, fir, and redwood in the United States, to say nothing of Spanish mahogany in British Honduras, African mahogany in West Africa, royal teak in Burma, or greenheart in British Guiana. In the age of the great interference the green daydreams of Constable, Wordsworth, and Cobbet, which evolved from an appreciation of country rides to an almost desperate sense of nature as a refuge from the industrial world, from man himself, disappeared. Disappeared, that is, from the arena of power, but not of the imagination. The imaginative daydream failed in the nineteenth century to realize itself in land policy or legislation – with a few exceptions, such as Yellowstone Park – and made itself felt primarily in literature, from the naturalism of Jack London to the Romantic historical fantasies of William Morris.[68]

A distinction can be made, however, between environmentalism as an intellectual phenomenon and environmentalism as legislative action and management of nature. Clearly Romanticism contributed heavily to a new understanding between humans and nature in the nineteenth century, and this strand of environmental thought constituted an important aspect of nineteenth- and early twentieth-century environmental thought. After the Second World War it has all but dominated the discourse. But in this earlier period the men who implemented the first forest reserves and launched conservationist environmentalism – Dalhousie, Brandis, Ribbentrop, Hough, and Pinchot – proclaimed a distinctly practical and utilitarian

---

[68] Yellowstone National Park, set aside in 1872, remains a notable exception. It served, however, to retard, not further the efforts of land reservation because the public and western senators did not want their state and federal land "locked up" in a similar manner.

gospel.[69] By escaping a Romantic "sentimentalism," as Fernow, chief of American forestry said, empire forestry gave the compromise from which environmentalism was born.[70]

The answer to the question – where and when did the environmental movement begin? – raises a theoretical question in its wake. The now classic essay by Lynn White, "The Historical Roots of our Ecological Crisis," asserts that Judeo-Christian society contains within itself hostility toward nature and is thus antiecological.[71] In the Middle Ages, he argues, Christian civilization posited a break between man and nature, and people ceased to think of themselves as a part of nature. They had the right to dominate earth. Though he recognized that western Christianity included differing accounts of the relationship between humans and the natural world, such as the pantheism of Saint Francis of Assisi, exploitation remained the dominant attitude. This attitude, according to White, merged with new technology and the industrial revolution to wreak havoc on the natural world.

Empire forestry undercuts this divorce of environmentalism from Christianity. Environmentalism paradoxically sprang from industrialism and imperialism and owes its origin to authoritarian, imperial, Christian, often racist Victorians of Anglo-Saxon and German heritage. Lynn White's assertion blaming Christian civilization for environmental woe appears to be, in the light of empire forestry, untrue.

The origins of environmentalism also undercut scholars who assert that environmentalism served as a "subversive rebel" to "imperial" scientific and political discourse. Like the ends of a horseshoe, scholars on the postmodernist left, who see environmentalism as abusive, and scholars on the right, who see European imperialism as the missionary of progress, are in fact closer together in this regard than the more comfortable and established academic left.[72] Clearly an authoritarian state working under the rule of western law gave birth to effective environmental policies. Whether this is a "good thing" or a "bad thing" has not been the purpose of this book to investigate. Many would agree with Ghandi's assessment of Brandis as "the hero of Pegu" who saved the Burmese forests, and many would not. Many

---

[69] Some success came for John Muir, nature writer and environmentalist, who represented the Romantic strain of environmentalism. His success in persuading the public and key officials to establish nature theme parks only retarded, however, the reservation of truly massive tracts of land as discussed earlier.

[70] Fernow, *Garden and Forest* 1 (1889): 371.

[71] Lynn White, Jr., "The Historical Roots of our Ecological Crisis," in *Machina ex Deo: Essays in the Dynamism of Western Culture* (Cambridge, MA, 1968).

[72] The scholarship of Nancy Peluso, Madhav Gadgil, Vandana Shiva, and K. Sivaramakrishnan upturn the assertion by Grove and Worster that environmental ideas were "progressive" and "subversive." Rather they show the construction of state power to be central to the environmental task. Sivaramakrishnan argues that the forestry policies in colonial India developed from a confluence of influence from John Stuart Mill, Lord Macaulay, and the evangelical movement, and resulted in a "pragmatic jurisprudence of suspicion, which favored surveillance, deterrence, and draconian measures for social control." See "Colonialism and Forestry," 11.

rightly point to the sense of loss that forest communities experience when officials foist controls and regulations on traditional culture – and the genuine suffering that such proscription entails.

The impact of environmentalism in the future may depend on the willingness of societies in every part of the globe to enforce a western view of property, private and state, and to enforce environmental law with effective police powers. The almost complete decay of the parks system in Africa and the destruction of forests in the Indian parks that once set the example for environmental resource management for the world, are a warning that western law and effective police power, once removed, may lead to environmental catastrophe.[73] Perhaps the utilitarian approach to nature – perfected by British imperialists – is best suited to balance the conflicting claims of romantic environmentalists, indigenous groups, farmers, industry, and urban society, and could reclaim for Africa and India the environmental gains they inherited from imperialism.

Admittedly a modernity based on bureaucratic overrationalization can overwhelm the individual and the community, upsetting the balance between state profit and local elites.[74] Even so, modern European civilization – including imperialism – must be credited with not only an industrial revolution that dramatically ended the Malthusian cycle in most of the world, but also gave birth to an environmental revolution which is still in the process of saving humans from themselves. The irony is that *this* green revolution, born of imperialism, has become a wildly popular appendage to the democratic movement. Interestingly, environmentalism injected into the democratic creed a decidedly hierarchical idea. To the egalitarian creed "the greatest good for the greatest number" has now been added, subversively, "the greatest good for the greatest number for the longest time." How a global consumer society will meet this environmental challenge remains to be seen. But one hopes the answer to the question – when and where did environmentalism begin? – will enable us to understand and perhaps perpetuate a revolution which, in magnitude of importance, takes its place alongside the Neolithic and industrial revolutions that preceded it.

[73] Recent scholarship has shown that short-term profit has recently been placed over the *sine qua non* of conservancy extracting more than the forest could replace. See R. Ali, S. N. Prasad and M. Gadgil, "Forest Management and Forest Policy in India: a Critical Review," *Social Action* 27 (1983): 1–50; R. Repetto and M. Gillis, *Public Policies and the Misuse of Forest Resources* (Durham, NC, 1989); N. S. Jodha, "Population Growth and the Decline of Common Property Resources in Rajasthan, India," *Population and Development Review* 11 (1985): 247–264.

[74] For an excellent analysis of the growth of both the state and individualism to the detriment of intermediate community groups, and its implication for community forestry, see Audun Sandberg, "Against the Wind: on Reintroducing Commons Law in Northern Norway," *Mountain Research and Development* 18 (1998): 95–106.

Bibliography

The following abbreviations have been used throughout the book:
JRGS   *Journal of the Royal Geographical Society*
JSA   *Journal of the Society of Arts*
JRSNSW   *Journal of the Royal Society of New South Wales*
PRCI   *Proceedings of the Royal Colonial Institute*

A. A. "Forestry in South Australia." *Indian Forester* 18, 1892.
"Administrative Report on the Railways in India." *Indian Forester* 18, 1892.
"Afforestation in England." *Indian Forester* 20, 1894.
"Afforestation in Hong Kong." *Indian Forester* 21, 1895.
"Afforestation, South Africa." *Indian Forester* 34, 1908.
"African Bamboo." *Indian Forester* 21, 1895.
"African Mahogany." *Indian Forester* 21, 1895.
"African Timber." *JSA* 63, 1910.
"The Aims and Future of Forest Research in India." *Indian Forester* 33, 1907.
Albion, G. *Forests and Sea Power: the Timber Problem of the Royal Navy , 1652–1862.*
    Cambridge, MA, 1926.
Alexander, Scott. *Locke.* New York, 1969.
Ali, R., Prasad, S. N., and Gadgil, M. "Forest Management and Forest Policy in India: a
    Critical Review." *Social Action* 27, 1983.
Allsop, F. *The First Fifty Years of New Zealand's Forest Service.* Wellington, 1973.
"American Bureau of Foresters in the Philippines." *Indian Forester* 30, 1904.
"The American Bureau of Forestry." *Indian Forester* 29, 1903.
"An American Working Plan." *Indian Forester* 20, 1894.
Anderson, Mark. *A History of Scottish Forestry.* 2 vols. Nelson, 1967.
Andrew, W. P. *Our Scientific Frontier.* London, 1880.
Armstrong, John A. *The European Administrative Elite.* Princeton, 1973.
Arneil, Barbara. "The Wild Indian's Venison: Locke's Theory of Property and English
    Colonialism in America." *Political Studies* 44:1, March 1996.
Arnold, Matthew. *Culture and Anarchy.* New York, 1941.
    "Dover Beach." *The Poems of Matthew Arnold.* London, 1913.
Arrian. *Anabasis and Indica.* Trans. E. J. Chinnock. London, 1893.

A. S. "Annual Progress Report of State Forest Administration in New South Wales for 1890." *Indian Forester* 18, 1892.

Annual Progress Report of State Forestry Administration in New South Wales for 1891. *Indian Forester* 19, 1893.

"Descriptive List of the Timber Trees of the C.P. of Ceylon." *Indian Forester* 18, 1892.

"At the St. Louis Fair." *Indian Forester* 30, 1904.

Aubréville, A. "Les forêts de la colonie du Niger." *Bulletin du Comité d'études Historiques et Scientifiques de l'Africque Occidentale Française* 19 (1936).

"Australian Forestry." *Indian Forester* 31, 1905.

"Australian Forestry." *JSA* 53, 1905.

Bacon, Francis. "The Great Instauration." In *The Works of Francis Bacon*, ed. James Spedding. 14 vols. New York, 1872–1878.

Baden Powell, B. H., and Macdonell, J. C. *Report of the Proceedings of a Conference of Forest Officers Held at Lahore, January 2 and 3, 1872*. Lahore, 1872.

Bailey, F. "Forestry in France." *Indian Forester* 15, 1886.

Baird, J. G. A. *Private Letters of the Marquis of Dalhousie*. Edinburgh and London, 1910.

Baker, Mark J. "The Effect of Community Structure on Social Forestry Outcomes: Insights from Chota, Nagpur, India." *Mountain Research and Development* 18, 1998.

Balfour, Edward. "The Forestry and Woods of Southern India." *JSA* 2, 1854.

   The Cyclopedia of India and of Eastern and Southern Asia Commercial, Industrial, and Scientific Products of the Mineral, Vegetable, and Animal Kingdoms, Useful Arts and Manufactures. London, 1885.

"Bamboos: their Study, Culture and Use." *Indian Forester* 33, 1907.

Bayly, C. A. "The Middle East and Asia During the Age of Revolutions, 1760–1830." *Itinerario* 2, 1986.

Behre, Karl-Ernst. "The Role of Man in European Vegetation History." In *Vegetation History*, ed. B. Huntley and T. Webb III. Dordrecht, 1988.

Berkes, Fikret, Davidson-Hunt, Ian, and Davidson-Hunt, Kerril. "Diversity of Common Property Resource Use and Diversity of Social Interests in the Western Indian Himalayas." *Mountain Research and Development* 18, 1998.

Berman, M. *Social Change and Scientific Organization, 1799–1844*. London, 1978.

Bermingham, Ann. *Landscape and Ideology: the English Rustic Tradition, 1740–1860*. Berkeley, 1986.

BHBP. "Protected Forests." *Indian Forester* 19, 1893.

Biddulph, C. E. *Our Western Frontier of India*. London, 1887.

Bidie, George. "Effects of Forest Destruction in Coorg." *JRGS* 39, 1869.

Bilsky, L. J., ed. *Historical Ecology: Essays on Environment and Social Change*. Port Washington, NY, 1980.

Birdwood, H. M. "The Hill Forests of Western India." *JSA* 47, 1898.

"Botanical and Afforestation Department of Hong Kong." *Indian Forester* 18, 1892.

"Botany and the Forest Department." *Indian Forester* 26, 1900.

Bowler, Peter, J. *The Environmental Sciences*. London, 1992.

Bramwell, F. J., and Wood, Trueman H. "Education in Forestry." *JSA* 30, 1882.

Brand, F. T. *Proposed Anti-Desiccation Scheme for Northern Nigeria*. Kaduna, 1936.

Brandis, Dietrich. "Pioneers of Indian Forestry: Dr. Hugh Cleghorn's Services to Indian Forestry." *Indian Forester* 31, 1905.

Brandis, D., and Smythies, A., eds. *Report of the Proceedings of the Forest Conference Held at Simla, October 1875.* Calcutta, 1876.

Brantlinger, Patrick. Editor. *Energy and Entropy: Science and Culture in Victorian Britain.* Bloomington, 1989.

Bricogne, M. "Les forêts de l'empire Ottoman." *Revue des Eaux et Forêts* 16, 1877.

Brincken, Baronde. *Mémoire descriptif sur la fôret impériale de Bialowieza en Lithuanie.* Chez Glucksberg, 1826.

"The British Association at Southport." *Indian Forester* 30, 1904.

British Empire Forestry Conference. London, 1921.

"British Empire Naturalists Association." *Indian Forester* 33, 1907.

Brown, John Croumbie. *French Forest Ordinance of 1669 with Historical Sketch of Previous Treatment of Forests in France.* Edinburgh, 1883.

Brown, J. Ednie. "The Forest of Western Australia." *Indian Forester* 26, 1900.

Bruce, C. W. A. "The Reorganization of the Imperial and Provincial Services." *Indian Forester* 21, 1895.

Buffon, Comte de. "De la nature. Première vue." *Histoire naturelle, générale et particulière.* vol. XII. Paris, 1774–1804.

Bureau of the Census. US Department of the Interior. *Compendium of the Eleventh Census, 1890.* Pt. 1 at 48, 1892.

"The Burma Forest Administration Report, 1896–97." *Indian Forester* 24, 1898.

Burn-Murdoch, A. M. "Notes from the Federated Malay States." *Indian Forester* 30, 1904.

Burnett, Gilbert. *Timber Trees of the Territory of Papua: Reports and Catalogue.* Melbourne, 1908.

Bury, J. B. *The Idea of Progress.* London, 1920.

Cameron, Rondo E. "The Credit Mobilier and the Economic Development of Europe." *Journal of Political Economy* 61, 1953.

Campbell-Walker, Ian. *Report of the Forest Department, Madras Presidency, for the Year 1885–86.* Dehra Dun, 1887.

Report of the Conservator of State Forests, with Proposals for the Organization and Working of the State Forest Department. Wellington, 1877.

Reports on Forest Management in Germany, Austria, and Great Britain. London, 1873.

"Canadian Forestry Journal, for March 1907." *Indian Forester* 33, 1907.

"Canadian Forests and Forestry." *Indian Forester* 52, 1904.

Cannon, S. F. *Science in Culture: the Early Victorian Period.* London, 1978.

"The Cape Forest Reports for 1892." *Indian Forester* 20, 1894.

"The Cape of Good Hope Forester Report for 1898." *Indian Forester* 26, 1900.

"The Cape of Good Hope, Report on Forest Administration for the Year 1904." *Indian Forester* 30, 1904.

Careter, Paul. *The Road to Botany Bay: an Exploration of Landscape and History.* Chicago, 1987.

Carron, L. T. *A History of Forestry in Australia.* Rushcutters Bay, New South Wales, 1985.

"Cause and Effect of the Gradual Disappearance of Forest on the Earth's Surface." *Indian Forester* 33, 1907.

"The Cedar of Central Africa." *Indian Forester* 21, 1895.

"The Ceylon Forest Administration Report for 1893." *Indian Forester* 21, 1895.

"Ceylon Forest Report for 1901." *Indian Forester* 29, 1902.

C. G. R. "Recruitment of Officers for the Indian Forest Service." *Indian Forester* 21, 1895.
    "Recruits for the Upper Controlling Staff of the Forest Department." *Indian Forester*
    20, 1894.

Chapais, J. C. *Canadian Forests: Illustrated Guide.* Montreal, 1885.

"Charcoal for Iron Smelting." *Indian Forester* 19, 1893.

"The Chicago Exhibition." *Indian Forester* 20, 1894.

"China." *JSA* 66, 1918.

Christie, W. "The Forest Vegetation of Central and Northern New England in Connection
    with Geological Influence." *JRSNSW* 11, 1878.

Cittandino, Eugene. *Nature as the Laboratory: Darwinian Plant Ecology in the German
    Empire, 1880–1900.* Cambridge, 1990.

Clark, C. N. H. *History of Australia.* 2 vols. Cambridge, 1962–1968.

Clarke, W. B. "Effects of Forests on Climate." *JRSNSW* 10, 1876.

Cohn, Gustav. *Zur Geschichte und Politik des Verkehrswesens.* Stuttgart, 1900.

"Colonial Forestry." *JSA* 24, 1886.

Colonial Reports Miscellaneous, no. 2. *Report on the Forests of Zululand.* London,
    1891.

Colonial Reports Miscellaneous, no. 51. *Southern Nigeria: Report on the Forest
    Administration of Southern Nigeria for 1906.* London, 1908.

Colony of Natal: Report on Forestry in Natal and Zululand. Pietermaritzburg, 1902.

"The Condition of Forestry in New Zealand." *Indian Forester* 18, 1892.

The *Control Journal of the Mudumalai Leased Forests and the Benne Reserved Forest.* Ed.
    J. H. Longrigg. Madras, 1924.

"Coppicing of Unprotected Forests." *Indian Forester* 26, 1900.

Corbin, H. Hugh. *Facts and Figures: Forestry in Australia, with Special Reference to South
    Australia.* Adelaide, 1913.

Cox, Charles. *The Royal Forests of England.* London, 1905.

Cronon, William. *Nature's Metropolis: Chicago and the Great West.* New York, 1991.

"The Cutting and Upkeep of Boundary Lines." *Indian Forester* 26, 1900.

Cyclopaedia of India and of Eastern and Southern Asia. Third edition. Madras, 1857.

Dalhousie, James Andrew Broun Ramsay. *Minute on Forest Policy.* August 3, 1855.
    Parliamentary Papers, Fort William.

    Private Letters of the Marquess of Dalhousie. Ed. J. G. A. Baird. London, 1911.

"The Danger of the Formation of the Pure Forest in India." *Indian Forester* 33, 1907.

Danvers, Julian. Commonwealth Institute, *Report on the Railways of India,* 1861–1862,
    1862–1863, 1863–1864, 1865–1866, 1866–1867.

Darby, H. C. *An Historical Geography of England Before AD 1800.* Cambridge, 1951.

Dargavel, John. *Fashioning Australia's Forests.* Melbourne, 1995.

Darian, S. *The Ganges in Myth and History.* Honolulu, 1978.

Daubeny, Charles. *Essay on the Trees and Shrubs of the Ancients.* Oxford and London, 1865.
    Lectures on Roman Husbandry. Oxford, 1857.

"Deforestation in Russia." *Indian Forester* 21, 1895.

"Destruction of Game in the C.P." *Indian Forester* 27, 1901.

"Destruction of Greek Forests." *Indian Forester* 28, 1902.

"The Devastation of the Forest in West Africa and the Diminution in the Water Supply."
    Indian Forester 31, 1905.

Drayton, Richard Harry. "Imperial Science and a Scientific Empire: Kew Gardens and the Uses of Nature, 1772–1903." Ph.D. thesis, Yale University, 1993.

Drew, J. *India and the Romantic Imagination*. Oxford, 1987.

Drummond, A. T. *Forest Preservation in Canada*. Montreal, 1885. Printed as an addendum to the *Report of the Annual Meeting at Boston of the America Forestry Congress*. Montreal, 1885.

Eardley-Wilmot, Saint-Hill. "Indian State Forestry." *JSA* 58, 1910.

East India (Retrenchment Committee). *Report of the Indian Retrenchment Committee 1922–23*. London, 1923.

Ebermayer, Ernst Wilhelm Ferdinand. Kremery Reference Files, biographical materials, folder 1900, 9999. University of Wisconsin, Madison.

"Edinburgh Forestry Exhibition." *JSA* 32, 1884.

Edney, Matthew. "Mapping and Empire: British Trigonometrical Surveys in India and the European Concept of Systematic Survey, 1799–1843." Ph.D. thesis, University of Wisconsin, Madison, 1990.

"The Effects of Forestry on the Circulation of Water at the Surface of Continents." *Indian Forester* 28, 1902.

"The Effects of Grazing on Forests." *Indian Forester* 26, 1900.

Ellenborough, Edward Law. *History of the Indian Administration of Lord Ellenborough: in his Correspondence with the Duke of Wellington: to which is Prefixed, by Permission of Her Majesty, Lord Ellenborough's Letters to the Queen During that Period. Edited by Reginald Colchester.* London, 1874.

Ellis, L. MacIntosh. *The Progress of Forestry in New Zealand. Prepared on the Occasion of the Sixteenth Meeting of the Australasian Association for the Advancement of Science.* Wellington, 1922.

Emery, F. V. "Geography and Imperialism: the Role of Sir Bartle Frere, 1815–1884." *Geographical Journal*, 1984.

Encyclopedia Britannica. Eleventh edn. Cambridge, 1910.

Erkkil, Antti, and Siiskonen, Harri. *Forestry in Namibia, 1850–1990*. Joensuu, 1992.

"Eucalyptus Screens as Fire Protection Belts." *Indian Forester* 31, 1905.

Evans, David. *A History of Nature Conservation in Britain*. London, 1992.

Evans, E. "Ethical Relations Between Man and Beast." *Popular Science Monthly*, September 1894.

Evelyn, John. *Silva, or a Discourse of Forest-Trees, and the Propagation of Timber in His Majesty's Dominions*. London, 1706.

Everest, George. *A Series of Letters Addressed to His Royal Highness the Duke of Sussex*. London, 1839.

Explanatory Memoranda for the Central Board of Forestry, 7, 8, 9 May 1951, Forest Research Institute and Colleges. Dehra Dun, 1951.

"Extermination of Wild Beasts in the Central Provinces." *Indian Forester* 18, 1892.

"Extraordinary Flights of Butterflies." *Indian Forester* 26, 1900.

F. G. "Forestry at the Cape." *Indian Forester* 25, 1899.

Fairhead, James, and Leach, Melissa. *Misreading the African Landscape: Society and Ecology in a Forest Savanna Mosaic*. Cambridge, 1996.

Faulkner, Robert K. *Francis Bacon and the Project of Progress*. Lanham, 1993.

Fernow, Bernard. *A Brief History of Forestry in Europe, the United States, and Other Countries*. Toronto, 1913.

Garden and Forest 1, 1889.

F. G. "Forestry at the Cape." *Indian Forester* 25, 1899.

Fisher, W. R. "An American Primer of Forestry." *JSA* 54, 1905–1906.

Fishwick, R. W. *Some Notes on the History of Forestry in Northern Nigeria*. 1961.

"The Food of Nestling Birds." *Indian Forester* 28, 1902.

"For Little Known Trees." *Indian Forester* 22, 1896.

"Forest Administration and Revenue Making." *Indian Forester* 31, 1905.

"Forest Administration in Kashmir, from 1891–1895." *Indian Forester* 22, 1896.

"Forest Administration in South Australia, 1893–1894." *Indian Forester* 21, 1895.

"Forest Administration in South Australia, 1898–99." *Indian Forester* 26, 1900.

Forest and Conservation History. Durham, NC, 1957.

"The Forest and Fauna of British Central Africa." *Indian Forester* 22, 1896.

"Forest and Mineral Wealth of Brazil." *JSA* 39, 1891.

"Forest Conservancy in Victoria." *Indian Forester* 18, 1892.

"Forest Conservation in New South Wales, 1899." *Indian Forester* 27, 1901.

"Forest Destruction and the Russian Famine." *Indian Forester* 18, 1892.

"Forest Distribution of the United States." *JSA* 34, 1886.

"Forest Fires." *Indian Forester* 19, 1893.

"Forest Fires in New Jersey." *JSA* 44, 1896.

"The Forest of the Ivory Coasts." *Indian Forester* 33, 1907.

"The Forest of Uganda." *Indian Forester* 28, 1902.

"The Forest of Zululand." *JSA* 39, 1891.

"Forest Offenses and their Prevention." *Indian Forester* 33, 1907.

"Forest Officers as Photographers." *Indian Forester* 22, 1896.

"Forest Products of British Guiana." *JSA* 41, 1893.

"The Forest Products of Madagascar." *JSA* 39, 1891.

"Forest Products of Siam." *JSA* 32, 1880.

"On Forest Settlement and Administration." *Indian Forester* 19, 1893.

"Forest Trees of Nicaragua." *JSA* 42, 1894.

"Forest Wealth of British Columbia." *JSA* 47, 1898.

"The Forest Wealth of New South Wales." *JSA* 44, 1896.

"Forest Work in South Australia, 1895–1897." *Indian Forester* 24, 1898.

Forester. "Conservancy of Forests." *Cape Monthly Magazine* 16, 1897.

"The Foresters of the United States." *Indian Forester* 22, 1896.

"Forestry and Water Supply." *Indian Forester* 28, 1902.

"Forestry at the Cape of Good Hope During 1899." *Indian Forester* 28, 1902.

"Forestry at the World's Fair." *Indian Forester* 31, 1905.

"Forestry: Australia." Reprinted from *The Official Year Book of the Commonwealth of Australia* 50, 1964.

"Forestry in America." *Indian Forester* 29, 1903.

Forestry in Britain: Reconstruction Committee's Report and its Applicability to New Zealand. Wellington, 1919.

"Forestry in Canada." *Indian Forester* 31, 1905.

"Forestry in German East Africa." *Indian Forester* 28, 1902.

"Forestry in Indochina." *Indian Forester* 31, 1905.

"Forestry in Madagascar." *Indian Forester* 25, 1899.

"Forestry in Madras." *JSA* 49, 1901.

"Forestry in New South Wales." *Indian Forester* 25, 1899.

"Forestry in New South Wales." *JSA* 59, 1911.

"Forestry in South Australia, 1891–92." *Indian Forester* 19, 1893.

"Forestry in the Hawaiian Islands." *Indian Forester* 30, 1904.

"Forestry in the Soudan." *Indian Forester* 29, 1900.

"Forestry Work in New Zealand." *JSA* 66, 1918.

"Forests and Famine in Bombay." *Indian Forester* 27, 1901.

"Forests Considered in their Relation to Rainfall and the Conservation of Moisture." *JRSNSW* 36, 1902.

"The Forests of Asia Minor." *JSA* 56, 1908.

"Forests of Chile." *Indian Forester* 31, 1905.

"The Forests of Finland." *JSA* 27, 1879.

"The Forests of Russia." *Indian Forester* 28, 1902.

"The Forests of Tasmania." *JSA* 39, 1891.

"The Forests of the Planet Mars." *Indian Forester* 33, 1907.

"Forests of the United States of North America." *Indian Forester* 19, 1893.

"The Forests of Tunis." *JSA* 34, 1886.

Foster, F. W. *Forestry in Relation to Soil and Water Conservation*. New Zealand Institute of Foresters, 1954.

"Franklin B. Hough." *American Forests and Forestry Life*, July 1922.

Futa. "On Forest Settlement and Administration." *Indian Forester* 19, 1893.

"The Future Training of the Upper Controlling Staff of the Service." *Indian Forester* 31, 1905.

Gadgil, Madhav., and Guha, Ramachandra. *This Fissured Land: an Ecological History of India*. New Delhi, 1992.

Gamble, J. S. "The Advantage of Preliminary Practical Work in the Training of Forest Officers." *Indian Forester* 18, 1892.

"Forestry at the Paris Exhibition of 1900." *Indian Forester* 27, 1901.

"Instruction in Forestry at Coopers Hill." *Indian Forester* 18, 1892.

Manual of Indian Timber. Calcutta, 1902.

Garnett, Richard. "The British Museum Catalogue as the Basis of a Universal Catalogue." *Essays in Librarianship and Bibliography*. London, 1899.

Gauthier, David. *Perspectives on Thomas Hobbes*. New York, 1988.

Gem. "A Plea for Protected Forests." *Indian Forester* 19, 1893.

"Protected Forests." *Indian Forester* 20, 1894.

"Geology." *Victorian Studies* 25, 1982.

Gibson, Alexander. "A General Sketch of the Province of Guzerat, from Deesa to Daman." *Transactions of the Medical and Physical Society of Bombay* 1, 1838.

Letters to John Hooker. March 24, 1841. India Letters. Royal Botanic Gardens Archives, Kew, Surrey.

Gibson, J. M. *Report of the Royal Commission on Forestry Protection in Ontario*. Toronto, 1899.

Gifford, John. "Forest Fires in New Jersey." *JSA* 44, 1896.

Gille, Bertrand. "Le moulin à eau." *Techniques et Civilisations* 3, 1954.

Gillis, Peter R., and Roach, Thomas. "Early European and North American Forestry in Canada: the Ontario Example, 1890–1941." In *History of Sustained-Yield Forestry: a Symposium*, ed. Harold K. Steen, Santa Cruz, CA, 1984.

  Lost Initiatives: Canada's Forest Industries, Forest Policy and Forest Conservation. Westport, CN, 1986.

Glacken, Clarence J. *Traces on the Rhodian Shore: Nature and Culture in Western Thought from Ancient Times to the End of the Eighteenth Century.* Berkeley, 1967.

Gleadow, F. "Forest Fires." *Indian Forester* 29, 1903.

Glick, Thomas F., and Kohn, David. *Darwin on Evolution: the Development of the Theory of Natural Selection.* Indianapolis, 1996.

Gole, Susan. *India Within the Ganges.* New Delhi, 1983.

  Indian Maps and Plans: from Earliest Times to the Advent of European Surveys. New Delhi, 1989.

Gottlieb, Robert. *Forcing the Spring: the Transformation of the American Environmental Movement.* Washington, DC, 1993.

Government of India (n.d.) Department of Environment: a Profile. Government of India, New Delhi.

"Gradual Effect of the Disappearance of the Earth's Forest." *Indian Forester* 33, 1907.

Grand, Roger, and Raymond Delatouche. *L'agriculture au moyen âge, de la fin de l'empire Romain au 16e Siècle.* Paris, 1950.

"Grazing and Commutation in the C.P." *Indian Forester* 18, 1892.

"Grazing in Forest Lands." *Indian Forester* 26, 1900.

"The Great Laboratory." *Indian Forester* 33, 1907.

Grieg, G. "Threatened Destruction of Forests in the North-West Provinces of India: its Causes and Consequences." *JRGS* 1, 1879.

Griffish, Grosvenor. *Population Problems in the Age of Malthus.* Cambridge, 1926.

Grove, Richard. *Green Imperialism: Colonial Expansion, Tropical Island Edens and the Origins of Environmentalism, 1600–1860.* Cambridge, 1995.

  "The Origins of Environmentalism." *Nature* 3, May 1990.

  "The Origins of Western Environmentalism." *Scientific American* 267, 1992.

Grut, Mikael. *Forestry and Forest Industry in South Africa.* A. A. Balkema/Cape Town/Amsterdam, 1965.

Guha, R. *The Unquiet Woods: Ecological Change and Peasant Resistance in the Himalaya.* New Delhi, 1989.

Gupta, Krishna Murti, and Bandhu, Desh, eds. *Man and Forests (a New Dimension in the Himalaya): Proceedings of the Seminars Held in Shillong, Dehra Dun and New Delhi and Organized by Himalaya Seva Sangh, Rajghat, New Delhi.* New Delhi, 1979.

Habermas, Jürgen. *The Structural Transformation of the Public Sphere.* Cambridge, 1989.

"The Habitat of the Red Junglefowl." *Indian Forester* 28, 1902.

Hagberg, Knut. *Carl Linnaeus.* London, 1952.

Hall, D. G. E. *The Dalhousie–Phayre Correspondence 1852–1856.* London, 1932.

Hamilton, A. C., and Bensted-Smith, R. *Forest Conservation in the Wast Usambara Mountains: Tanzania.* Tanzania, 1989.

Hanna, H. B. "India's Scientific Frontier, Where is It? What is It?" *India's Problems* 2, 1895.

"The Harmfulness of Bushfires in the West Indies." *Indian Forester* 27, 1901.

Hawthorne, W. D., and Abu-Juam, M. *Forest Protection in Ghana*. Ghana, 1996.

Hays, Samuel. *Conservation and the Gospel of Efficiency: the Progressive Conservation Movement 1890–1920*. New York, 1959.

Headrick, D. R. *Tools of Empire*. New York, 1981.

Heidegger, Martin. "The Age of the World Picture." In *The Question Concerning Technology and Other Essays*. New York, 1977.

Herder, Johann. *Outlines of a Philosophy of the History of Man*. Trans. T. Churchill. London, 1800.

Heske, Franz. *German Forestry*. New Haven, 1938.

Heyck, Thomas William. *The Peoples of the British Isles: a New History From 1688 to 1870*. Belmont, CA, 1992.

The Peoples of the British Isles: a New History From 1870 to the Present. Belmont, CA, 1992.

The Transformation of Intellectual Life in Victorian England. New York, 1982.

Heyne, Moriz. *Das deutsche Nahrungswesen von den Ältesten Geschichtlichen Zeiten bis zum 16. Jahrhundert*. Leipzig, 1901.

"The History of a Railway Sleeper." *Indian Forester* 18, 1892.

Hobson, J. A. *Imperialism: a Study*. Ann Arbor, 1965.

Holdich, T. H. *The Indian Borderland, 1880–1900*. London, 1909.

Political Frontiers and Boundary Making. London, 1916.

Hooker, W. J. *Exotic Flora Containing Figures & Descriptions of New, Rare, & Otherwise Interesting Exotic Plants, Especially of such as are Deserving of being Cultivated in our Gardens*. Edinburgh, 1823–1827.

Hopkins, W. *The Great Epic of India*. New York, 1902.

Hoskins, W. G. *The Making of the English Landscape*. Harmondsworth, 1955.

Hoskins, W. G., and Finberg, H. R. *Common Lands of England and Wales*. London, 1963.

Hough, Franklin B. *Diaries*. Nov. 20, 1873, Feb. 2–12, 1874.

On the Duty of Governments in the Preservation of Forests. Salem, 1873.

Papers. New York State Library, Albany.

Report Upon Forestry, 4 vols. Washington, DC, 1882.

H. S. "Too Much Fire-Protection in Burma." *Indian Forester* 22, 1896.

Huffel, G. *Economie forestière*. 3 vols. Paris, 1904–1907.

Humboldt, Alexander. *Edeen zu einer Geographie der Pflanzen nebst einem Naturgemälde der Tropenländer*. Tübingen, 1807.

Essai politique sur le royaume de la nouvelle-Espagne. Paris, 1811.

Humboldt, Alexander, and Aimé Bonpland. *Nova genera et species plantarum*. Weinheim, 1815.

Voyage aux régions equinoxiales du nouveau continent 1790–1804. Paris, 1814, 1816.

Hume, David. *Hume Selections*. Ed. Charles W. Hendel, Jr. New York, 1927.

Huntley, B., and T. Webb III, eds. *Vegetation History*. Kluwer Academic Publishers, 1988.

Hunter, W. W. *Famine Aspects of the Bengal Districts*. Simla, 1873.

Hutchins, D. E. "The Cluster-Pine in South Africa." *Indian Forester* 24, 1894.

A Discussion of Australian Forestry, with Special References to Forestry in Western Australia, the Necessity of an Australian Forestry Policy, and Notices of Organised Forestry in Other Parts of the World Together with Appendices Relating to Forestry in New Zealand, Forestry in South Africa, and Control of the Rabbit Pest. Perth, 1916.

"The Forest of Natal." *Indian Forester* 13, 1892.
"An Invitation to Indian Foresters in South Africa." *Indian Forester* 26, 1900.
Report on Cyprus Forestry. London, 1909.
Report on the Forests of British East Africa. London, 1909.
Transvaal Forest Report. Pretoria, 1903.
Waipoua Kauri Forest, its Demarcation and Management. Wellington, 1918.
Hutton, James. *James Hutton's System of the Earth, 1785; Theory of the Earth, 1788; Observations on Granite, 1794.* New York, 1970.
"Improvement Fellings." *Indian Forester* 33, 1907.
"India and Australia." *Indian Forester* 21, 1895.
"India Rubber." *Indian Forester* 22, 1896.
"The Influence of Forestry on Water Supply." *Indian Forester* 18, 1892.
"The Influence on the Vegetation of a Forest of the Removal of Dead Leaves from the Soil." *Indian Forester* 19, 1893.
"The Insect World in an Indian Forest and How to Study It." *Indian Forester* 28, 1902.
"Insufficiency of the World's Timber Supply." *Indian Forester* 27, 1901.
James, N. D. G. *A History of English Forestry.* Oxford, 1981.
Jerram, M. R. K. *Report on the Group System of Natural Regeneration in Germany and its Application to Indian Forests.* Simla, 1913.
Jevons, William Stanley. *The Coal Question: an Inquiry Concerning the Progress of the Nation and the Probable Exhaustion of our Coal Mines.* London, 1866.
Jodha, N. S. "Population Growth and the Decline of Common Property Resources in Rajasthan, India." *Population and Development Review* 11, 1985.
Journal of Arnold Arboretum. April 1927.
Jurisprudence générale. Répertoire méthodique et alphabétique de législation de doctrine et de jurisprudence. Ed. D. Dalloz and Armand Dalloz. Paris, 1849.
Kala, J. C. "People's Participation in Public Wastelands Development in India: an Analysis." *Indian Forester* 120, 1994.
Kamper, Dietmar, and Wulf, Christoph, eds. *Looking Back on the End of the World.* New York, 1989.
Kant, Immanuel. *Critique of Pure Reason.* Trans. F. Max Müller. New York, 1902.
Kensington, W. C. *New Zealand Department of Lands Report on Scenery Preservation for the Year 1908–1909.* Wellington, 1909.
New Zealand *Department of Lands Forestry in New Zealand.* Wellington, 1909.
Kirk, T. *Native Forests and the State of the Timber Trade. New Zealand.* 1886.
Kirkwood, Alexander. *Papers and Reports upon Forestry, Forest Schools, Forest Administration and Management in Europe, America, and the British Possessions, and upon Forests as Public Parks and Sanitary Resorts to Accompany the Report of the Royal Commission on Forest Reservation and National Parks.* Toronto, 1893. National Agricultural Library.
Kipling, Rudyard. *The Jungle Book.* Oxford, 1987.
Koroleff, A. "Fundamental Cause of our Failure in Forestry and the Remedy." In *Report for the Forestry Research Conference at the National Research Council of Canada.* Ottowa, 1935. Printed in *Woodlands Reviews* 6, 1935.
Kotey, Nii Ashie, *et al. Falling Into Place.* London, International Institute for Environment and Development; Legon, Ghana, Ministry of Lands and Forestry, 1998.

Lacey, John F. "The Destruction and Repair of Natural Resources in America." *Indian Forester* 22, 1896.

Lambton, William. "An Account of the Measurement of an Arc on the Meridian on the Coast of Coromandel, and the Length of a Degree Deduced Therefrom in the Latitude 12[ring]32." *Asiatic Researches* 8, New Delhi, 1992.

Lane-Poole, C. E. *Assessment of 1935•85 Acres of Monterery Pine at Mount Burr Forest Reserve: Carried out by Students of the Australian Forestry School Under the Direction of C. E. Lane-Poole in August 1926.* Melbourne, 1927.

Report on the Forests of Norfolk Island. Victoria, n.d.

Lare-Prole, C. E., and Poole, A. L. *Nazi Influence on German Forest Administration.* New Zealand, 1947.

"Late Henry Baden-Powell, CIE." *Indian Forester* 27, 1901.

L. E. A. "Open Fire Lines in Coorg." *Indian Forester* 28, 1902.

Lecoy, A. *Suggestions on Forests in New Zealand.* Wellington, 1880.

Ledzion, Mary McDonald. *Forest Families.* London, 1991.

Lefebvre des Noëttess, Richard. *L'attelage, le cheval de selle à travers les âges. Contribution à l'histoire de l'esclavage.* Vol. I. Paris, 1931.

Lewis, N. B. *A Hundred Years of State Forestry: South Australia, 1875–1975.* Australia, 1975.

Leyton, L. "Soil Conditions and Tree Growth." *Chemistry and Industry*, 1947.

Libecap, Gary D. "Bureaucratic Opposition to the Assignment of Property Rights Over Grazing on the Western Range." *Journal of Economic History* 41, 1981.

Linnaeus, Carl. *Travels.* Ed. David Black. New York, 1979.

Longrigg, J. H., ed. *Control Journal of the Mudumalai Leased Forests and the Benne Reserved Forest.* Madras, 1924.

Lovat, Major-General. "Forestry." *JSA* 69, 1921.

Lowod, Henry E. "The Calculating Forester: Quantification, Cameral Science, and the Emergence of Scientific Forestry Management in Germany." In *The Quantifying Spirit in the Eighteenth Century*, ed. Tore Irängsmyr, J. L. Heilbron, and Robin E. Rider. Berkeley, 1990.

Lückhoff, H. A. "The Story of Forestry and its People." *In Our Green Heritage*, ed. W. F. E. Immelman, C. L. Wicht, and D. P. Actierman. Cape Town, SA, 1973.

Lushington, A. W. "The Necessity for Fire Protection." *Indian Forester* 30, 1904.

Lyell, Charles. *Principles of Geology.* Chicago, 1990.

Mabey, Richard. *Gilbert White: a Biography of the Author of the Natural History of Selborne.* London, 1986.

Mackay, D. *In the Wake of Cook: Exploration, Science, and Imperialism.* London, 1984.

Mackenzie, John, M., ed. *Imperialism and the Natural World.* Manchester, 1990.

MacLeod, R. "The Alkali Acts Administration, 1863–1884: the Emergence of a Civil Scientist." *Victorian Studies*, 1965.

"Of Medals and Men: a Reward System in Victorian Science, 1826–1914." *Notes and Records of the Royal Society* 26, 1971.

"Scientific Advice for British India: Imperial Perceptions and Administrative Goals, 1898–1923." *Modern Asian Studies* 9, 1975.

"On Visiting the Moving Metropolis: Reflections on the Architecture of Imperial Science." *Historical Records of Australian Science* 53, 1982.

MacMunn, George. *The Romance of the Frontiers.* Quetta, 1978.
Macpherson, Peter. "Some Causes of the Decay of the Australian Forests." *JRSNSW* 25, 1891.
"Madagascar Rubber." *Indian Forester* 26, 1900.
Magazine of Natural History, and Journal of Zoology, Botany, Mineralogy, Geology, and Meteorology. London, 1835.
Mahabharata. Trans. Pratap Chundra Roy. Calcutta, 1883–1894.
Maiden, J. H. "Forests Considered in their Relation to Rainfall and the Conservation of Moisture." *JRSNSW* 36, 1902.
"Marram Grass in Australia." *Indian Forester* 21, 1895.
"Where are the Largest Trees in the World?" *Indian Forester* 30, 1904.
Malin, J. C. *The Grassland of North America: Prolegomena to its History.* Gloucester, MA, 1947.
Markham, Clements R. "On the Effects of the Destruction of Forestry in the Western Ghats of India on the Water Supply." *JRGS* 36, 1866.
A Memoir on the Indian Surveys. London, 1871.
Marsh, George Perkins. "Agricultural Society of Rutland County." Sept. 30, 1847. Rutland, VT, 1848.
Man and Nature: Or, Physical Geography as Modified by Human Action. Cambridge, 1864/1965.
Maury, L. F. A. *Les forêts de la Gaule et de l'ancienne France.* Paris, 1867.
McCrimmon, Barbara. *Power, Politics, and Print: the Publication of the British Museum Catalogue, 1881–1900.* Hamden, CT, 1981.
McKelvey, Peter. *Steepland Forests: a Historical Perspective of Protection Forestry in New Zealand.* Canterbury, 1995.
Michael, J. "Forestry." *JSA* 43, 1894–1895.
"The Mines and Forests of Syria." *JSA* 52, 1904.
Modern Asian Studies 9, 1975.
"The Mongoose in the West Indies." *Indian Forester* 22, 1896.
Montesquieu, Charles. *Pensées et fragments inédits de Montesquieu.* Ed. Gaston de Montesquieu. Paris, 1899–1901.
Morgan, J. H. *Report on the Forests of Canada, in which is shewn the Pressing Necessity which Exists for their more Careful Preservation and Extension by Planting, as a Sure and Valuable Source of National Wealth.* Ottawa, 1896.
Mourisson, Felix. *Philosophies de la nature: Bacon, Boyle, Toland, Buffon.* Paris, 1887.
Munro, Thomas. "Timber Monopoly in Malabar and Canada." In *Major-General Sir Thomas Munro: Selections from his Official Minutes and Other Writings,* ed. A. J. Aburnoth. Vol. I. London, 1881.
Murchison, Sir Roderick Impey. *The Geology of Russia in Europe and the Ural Mountains.* London, 1845.
Muriel, C. F. "Forest Exploration in the Bahr-el-Ghazal (Sudan)." *Indian Forester* 29, 1903.
Myers, Greg. "Nineteenth-Century Popularizations of Thermodynamics and the Rhetoric of Social Prophecy." In *Energy and Entropy: Science and Culture in Victorian Britain,* ed. Patrick Brantlinger. Bloomington, 1989.
Nash, Richard. *The Rights of Nature: a History of Environmental Ethics.* Madison, 1989.
Wilderness and the American Mind. New Haven, 1967.

National Academy of Science: Biographical Memoirs. Vol. xii. London, 1929.

National Agricultural Library. Ministry of Lands. *Papers Relating to State Forests their Conservation, Planting, Management, & c.: Presented to Both Houses of the General Assembly.* Wellington, 1874.

National Archives, New Delhi. Miscellaneous Branch. *Consultations.* May 27, 1859, nos. 381–382, paras. 145–159.

Governor General's Camp Letters to the Court of Directors. No. 21. April 20, 1849.

Home Public Consultations. John Lawrence to the Secretary of State, Letter to Revenue-Forests, November 23, 1867, no. 24.

Home Public Consultations, Foreign and Political Select Committee Proceedings. No. 12. Jan. 6 to Dec. 29, 1767.

Letters From the Secret Committee. Sept. 6, 1852, no. 1524, para. 21. June 22, 1853, no. 23.

Letters to the Secret Committee. Jan. 5, 1853, no. 3. Aug. 8, 1855, no. 45.

Political Consultations. Dec. 28, 1855 (no. 319, paras. 2, 4, 10–27, 28–44). June 6, 1856 (no. 193, paras. 19–22, 47–56).

Political Letters from the Court of Directors and the Secretary of State. Nov. 21, 1855, no. 33, para. 2. Aug. 22, 1856, no. 83, paras. 10–18. Dec. 10, 1856, no. 47, paras. 2–3, 4.

Secret Consultations. 1849 (nos. 21, 41), 1852 (nos. 1, 4, 15), 1853 (nos. 11, 12, 71, 73), 1855 (no. 2).

"National Forestry." *Indian Forester* 26, 1900.

Negi, S. S. *Indian Forestry: 1947–1997.* Dehra Dun, 1998.

"The New Draft Rules Regarding Settlement and the Positions of Revenue and Forest Officers." *Indian Forester* 19, 1893.

Nigam, S. "The Making of a Colonial Stereotype – the Criminal Tribes and Castes of North India." Part 1 of "Disciplining and Policing the Criminals by Birth." *Indian Economic and Social History Review* 27:2, 1990.

"The Nilgilri Game and Fish Preservation Association." *Indian Forester* 31, 1905.

Nisbet, John. "Soil and Situation in Relation to Forest Growth." *Indian Forester* 20, 1894.

Norman-White, H. G. *Report on Certain Indigenous Timbers of India, Burma and the Andamans Considered Suitable for Railway Carriage-Building, 1924–25.* Lucknow, 1925.

"Notes from an American Forest Reserve." *Indian Forester* 29, 1903.

"Notice Regarding New South Wales Forestry Department." *JRSNSW* 25, 1891.

O. C. "Fire Conservancy." *Indian Forester* 25, 1899.

"Forestry in Japan." *JSA* 59, 1911.

"Punishments for Cattle Trespass." *Indian Forester* 20, 1894.

Office of the High Commissioner for India. *Report of the Indian Students' Department, 1922–23.* London, 1923.

Oliphant, F. M. *A Further Report on Forestry Development in Nigeria.* Nigeria, 1934.

Ondermas, J. A. C. "The Topography of the East Indies, and the Rise of the Utrecht Laboratory, 1850–1900." *Historical Science* 26, 1984.

"Open Fire Lines in the Coorg." *Indian Forester* 28, 1902.

Orders of the House of Assembly, June 4–5, 1888. *Report of the Select Committee on Forests Bill.* Capetown, 1888.

Osborne, M. "The Société zoologique d'acclimatation and the New French Empire: the Science and Political Economy of Economic Zoology During the Second Empire." Ph.D. thesis, University of Wisconsin, 1987.

"Outlook for Forestry in the Philippines." *Indian Forester* 28, 1902.

Ovington, J. D. *The Role of Forestry.* Canberra, 1965.

"Pacific Forests and Rainfall." *JSA* 33, 1885.

Padhi, G. S. *Forestry in India.* Dehra Dun, 1982.

Pant, G. B. *The Forest Problem in Kumaon.* Gyanodaya Prakashan, 1922.

"The Paris Exhibition." *Indian Forester* 25, 1899; 26, 1900.

Parliamentary Papers. The Punjab, 1847–1849. no. 53.

Pearson, G. F. "Recollection of the Early Days of the Indian Forest Department, 1858–1864." *Indian Forester* 29, 1903.

"The Teaching of Forestry." *JSA* 30, 1881–1882.

"In the Uganda Forests." *Indian Forester* 31, 1905.

Pearson, R. S. "The Recent Industrial and Economic Development of Indian Forest Products." *JSA* 65, 1917.

Peluso, N. L. "The History of State Forest Management in Java." *Forest and Conservation History* 35, 1991.

Perkin, Harold. *Origins of Modern English Society.* London, 1969.

The Rise of Professional Society. London 1989.

The Structured Crowd: Essays in English Social History. Sussex, 1981.

The Third Revolution: Professional Elites in the Modern World. London, 1996.

"The Perpetuation of Canadian Forests." *Indian Forester* 26, 1900.

Phayre, A. *Report on the Administration of the Province of Pegu for 1854–1855 and 1855–1856.* Calcutta, 1857.

Phillimore, R. H. *Historical Records of the Survey of India.* Dehra Dun, 1945.

Phipps, Robert W. *Report on the Necessity of Preserving and Replanting Forests: Compiled at the Instance of the Government of Ontario.* Toronto, 1883.

Pinchot, Gifford. *Breaking New Ground.* Washington, DC, 1947.

"The Forest of Ne-he-ha-sa-Ne Park in Northern New York." *Indian Forester* 19, 1893.

"Forestry Abroad and at Home." *National Geographic Magazine,* March 16, 1905.

"Government Forestry Abroad." *American Economic Association* 6:3, May 1891.

"A Primer of Forestry, Part 1, The Forest." *Direction of Forestry* 24, US Department of Agriculture, 1903.

Pinkett, Harold T. *Gifford Pinchot: Private and Public Forester.* Urbana, 1970.

"Pioneers of Indian Forestry: Colonel G. F. Pearson." *Indian Forester* 30, 1904.

"A Plea for Protected Forests." *Indian Forester* 19, 1893.

"Political Famine (Africa): a Plea for Forest Conservation." *Indian Forester* 27, 1901.

Ponting, Clive. *A Green History of the World: the Environment and the Collapse of Great Civilizations.* New York, 1991.

"President Roosevelt on Forestry." *Indian Forester* 29, 1903.

Prest, J. M. *The Garden of Eden.* London, 1981.

Probable Exhaustion of our Coal Mines. London, 1866.

"Probable Impact of Deforestation on Hydrological Processes." *Climate Change* 19, 1991.

Proceedings of the First Nigerian Forestry Conference: the Role of Forestry in the Economic Development of the Savanna Areas of Nigeria. Kaduna, Northern Nigeria, February 1964.

Proceedings of the First Senior Forest Officers Conference held at Dehra Dun from the 5th to the 7th April 1945. Simla, 1946.

Proceedings of the Second Senior Forest Officer's Conference held at Dehra Dun from the 7th to 9th November 1946. Simla, 1950.

Proceedings of the Sub-Committee. *Public Service Commission.* Simla, 1888.

"Progress Report of Forest Conservancy in Ceylon for 1891." *Indian Forester* 19, 1893.

"The Prohibition of Grass Burning and its Effect on the Game of the Country." *Indian Forester* 31, 1905.

"Proportionate Fellings in Selection Areas." *Indian Forester* 28, 1902.

"Proposed Forest Service in the Hawaiian Islands." *Indian Forester* 31, 1905.

"Punishments for Cattle Trespass." *Indian Forester* 20, 1894.

Pyenson, L. "Astronomy and Imperialism: J. A. C. Ondermas, the Topography of the East Indies, and the Rise of the Utrecht Laboratory, 1850–1900." *History of Science* 26, 1984.

"Cultural Imperialism and the Exact Sciences: German Expansion Overseas, 1900–1930." *History of Science* 20, 1982.

"*In partibus infidelium:* Imperialist Rivalries and Exact Sciences in Early Twentieth-Century Argentina."*Quina* 26, 1984.

Pyne, Stephen J. *Burning Bush: a Fire History of Australia.* New York, 1991.

Rackham, Oliver. *Hayley Wood, its History and Ecology.* Naturalists Trust, Cambridge, 1975.

The History of the Countryside. London, 1986.

Rahdar. "Influence of Places on Spirits." *Indian Forester* 20, 1894.

Rakestraw, L. "Conservation Historiography: an Assessment." *Pacific Historical Review* 41, 1972.

Ramayana. Translated by M. N. Dutt. Calcutta, 1894.

Raper, F. V. "Narrative of a Survey for the Purpose of Discovering the Source of the Ganges." *Asiatic Researches* 11, 1818/1979.

Rawat, Ajay, ed. *History of Forestry in India.* New Delhi, 1991.

"Re-Afforestation in Italy." *Indian Forester* 30, 1904.

Reconstruction Committee of the Council: Second Report on Reconstruction Planning. Delhi, 1944.

Records of the Royal Society 26, 1971.

"Regeneration of the Teak Forests of Java." *Indian Forester* 33, 1907.

Régnault, F. "Deboisement and Decadence." *Indian Forester* 30, 1904.

"Remembering Franklin B. Hough." *American Forests* 86, January 1977.

Rennell, James. *The Journals of Major James Rennell, First Surveyor-General of India, Written for the Information of the Governor of Bengal During his Surveys of the Ganges and Brahmaputra Rivers, 1764–1767.* Ed. T. H. D. LaTouche. Calcutta, 1910.

Rennie, Peter J. *Some Long-Term Effects of Tree Growth on Soil Productivity.* Chalk River, Ontario, 1961.

Repetto, R., and Gillis, M. *Public Policies and the Misuse of Forest Resources.* Durham, NC, 1989.

"Report on the Botanical and Afforestation Department, Hong Kong, For 1894." *Indian Forester* 21, 1895.

Report of Commission on Forest Reservation and National Park: and Papers and Reports upon Forestry, Forest Schools, etc. Toronto, 1893.

Report of the Commissioner of Crown Lands for Ontario. 1895.

Report of a Committee Appointed by the Secretary of State for the Colonies. *The Training of Candidates and Probationers for Appointment as Forest Officers in the Government Service.* Colonial no. 61. July 1931.

Report of a Committee Appointed by the Secretary of State for India to Enquire into the Recruitment and Training of Probationers for the India Forest Service. London, 1908.

Report on the Forest Wealth of Canada. Department of Agriculture, Statistical Office, Ottawa, 1895.

Report of the Forestry Committee. Simla, 1929.

Report of the National Commission on Agriculture, 1976, Part 9, Forestry. New Delhi, 1976.

"The Report of the New South Wales Department of Agriculture and Forestry, for 1895–97." *Indian Forester* 24, 1898.

Report on the Railways of India, 1862–67.

Report of the Royal Commission on Forest Reservation and National Park. Toronto, 1893.

Report of the Select Committee on Forests Bill. Cape Town, 1888.

Report on Timber used in Railway Carriage and Wagon Shops in India and Burma during 1934–35, 1935–36, and 1936–37. New Delhi, 1938.

"The Restoration of Scenery." *Indian Forester* 20, 1894.

"The Retirement of Mr. Berthold Ribbentrop, CIE." *Indian Forester* 26, 1900.

"The Retirement of Mr. J. S. Gamble, MA, FLS, from the Forest Service." *Indian Forester* 25, 1899.

Ribbentrop, Berthold. "The Forest of Victoria." *Indian Forester* 21, 1895.

Forestry in British India. Calcutta, 1900.

Reviews of Forest Administration in British India for the Year 1885–86. Dehra Dun, 1887.

Ricardo, David. *On the Principles of Political Economy.* London, 1992.

Richards, Thomas. *The Imperial Archive: Knowledge and the Fantasy of Empire.* London, 1993.

Richardson, E. P. *Painting in America: the Story of 450 Years.* New York, 1956.

Richardson, Elmo R. *The Politics of Conservation: Crusades and Controversies, 1897–1913.* Berkeley, 1962.

Robbins, William G. *American Forestry: a History of National, State, and Private Cooperation.* Lincoln, NB, and London, 1985.

Robinson, Roy. "Forestry in the British Empire." *JSA* 84, 1936.

Rosevear, D. R. *Vegetation, Forestry and Wild Life in Nigeria.* 1953.

Royal Commission on the Public Services in India. *Appendix to the Report of the Commissioners.* Vol. xv. London, 1915.

"Royal Society for the Protection of Birds." *Indian Forester* 33, 1907.

"Rules to Regulate Appointments and Promotions in the Provincial Forests Service." *JRGS* 13, 1892.

Sandberg, Audun. "Against the Wind: on Reintroducing Commons Law in Northern Norway". *Mountain Research and Development* 18, 1998.

Sargent, Charles. "Arbor Day." *Garden and Forest* 1, April 11, 1888.

"Forest Reservation in the United States." *Indian Forester* 23, 1897.

"The Future of our Forests." *Garden and Forest* 1, March 14, 1888.

"The Nation's Forests." *Garden and Forest* 1, January 30, 1889.

"Periodical Literature." *Garden and Forest* 1, March 21, 1888.

Garden and Forest 9, May 13, 1896.

Garden and Forest 9, May 27, 1896.

Garden and Forest 9, August 5, 1896.

Scammell, Edward T. "The Timber Resources of the Australian Commonwealth." *JSA* 50, 1902.

"The Timber Resources of the Australian Commonwealth." *Indian Forester* 28, 1902.

Schlater, W. L. "The Economic Importance of Birds in India." *Indian Forester* 18, 1892.

Schleiden, Mathias Jacob. *The Plant: a Biography.* Trans. Arthur Henfrey. London, 1853.

Schlich, W. "Forestry in the Colonies and in India." *PRCI* 21, 1889–1890.

Manual of Forestry: Forest Policy in the British Empire. Vol. I. London, 1906.

"Scientific Forestry." *Indian Forester* 33, 1907.

Schoepflin, F. *Colony of Natal: Report of the Conservator of Forests.* Pietermaritzburg, 1892.

Sears, Paul. "Ecology – a Subversive Subject." *BioScience* 14, July 1964.

Secord, J. "King of Siluria: Roderick Murchison and the Imperial Theme in Nineteenth-Century British Geology." *Victorian Studies* 25, 1982.

Session 1933–34. *Joint Committee on Indian Constitutional Reform.* Vol. I (part 1). London, 1934.

Sessional Paper no. 7 of 1934. *Report on the Commercial Possibilities and Development of the Forests of Nigeria.* Lagos, 1934.

Sessional Paper no. 6 of 1935. *A Further Report on the Commercial Possibilities and Development of the Forests in Nigeria.* Lagos, 1935.

Sessional Paper no. 37 of 1937. *Report of the Anglo-French Forestry Commission, 1936–37.* Lagos, 1937.

Sicherman, Barbara. *Alice Hamilton: a Life in Letters.* Cambridge, MA, 1984.

"Working it Out: Gender, Profession, and Reform in the Career of Alice Hamilton." In Noralee Frankel and Nancy S. Dye, eds., *Gender, Class, Race, and Reform in the Progressive Era.* Lexington, KT, 1991.

Simmonds, P. L. "Past, Present, and Future Sources of the Timber Supplies of Great Britain." *JSA* 33, 1885.

"Sir Dietrich Brandis." *Indian Forester* 33, 1907.

Sivaramakrishnan, K. "Colonialism and Forestry in India." *Comparative Studies in Society and History* 37:1, January 1995.

Sleeman, W. H. *Rambles and Recollections of an Indian Official.* Karachi, 1844/1973.

Spenglar, Oswald. *The Decline of the West.* New York, 1929.

Stafford, R. A. "Geological Surveys, Mineral Discoveries, and British Expansion, 1835–1871." *Journal of Imperial and Commonwealth History* 12, 1984.

"The Role of Sir Roderick Murchison in Promoting the Geographical and Geological Exploration of the British Empire and its Sphere of Influence, 1855–1871." D.Phil. thesis, University of Oxford, 1986.

"State Forest Administration in South Australia for 1899–1900." *Indian Forester* 27, 1901.

"State Forestry Administration in South Australia." *Indian Forester* 25, 1899.

Stearns, Raymond Phineas. *Science in the British Colonies of America*. London, 1970.

Stebbing, Edward. "Forestry and the War." *JSA* 64, 1916.

The Forests of India. 4 vols. New Delhi, 1982.

The Forests of West Africa and the Sahara: a Study of Modern Conditions. London and Edinburgh, 1937.

"Pioneers of Indian Forestry: Captain Forsyth and the Highlands of Central India." *Indian Forester* 30, 1904.

The Threat of the Sahara. *Geographical Journal* 85, 1935.

Steen, Harold. *The US Forest Service: a History*. Seattle, 1976.

Stenton, Doris. *English Society in the Early Middle Ages*. Harmondsworth, 1951.

Stokes, Eric. *The English Utilitarians in India*. Oxford, 1959.

Stone, Edward, ed. *What Was Naturalism?* Englewood Cliffs, NJ, 1959.

Storr-Lister, J. "Tree Planting in Punjab." *Cape Monthly Magazine* 14, 1877.

Strachey, R. "On the Physical Geography of the Provinces of Kumóan and Garhwal in the Himalaya Mountains, and of the Adjoining Parts of Tibet." *JRGS* 21, 1851.

"A Summary of Observed Results of Fire Protection in Reserved Forests." *Indian Forester* 31, 1905.

Swellendam. "Conservancy of Forest." *Cape Monthly Magazine* 16, 1878.

"Symbiosis and its Effects on the Planting of Forest Trees." *Indian Forester* 22, 1896.

Tautha. "A Plea for our Feathered Friends." *Indian Forester* 18, 1892.

Tayler, W. "Famines in India: their Remedy and Prevention." *East India Association* 7, 1873.

Tessier, L. F. "L'idée forestière dans l'histoire." *Revue des Eaux et Forêts*, January–February 1905.

Theal, G. M. *History of South Africa: 1798–1828*. London, 1903.

Thorner, Daniel. "Great Britain and the Development of India's Railways." *Journal of Economic History* 11, 1951.

Investment in Empire: British Railway and Steam-Shipping Enterprise in India, 1825–1849. Philadelphia, 1950.

Thomas, Keith. *Man and the Natural World: a History of the Modern Sensibility*. New York, 1983.

Thompson, H. N. *Report Regarding the Irregularities of Rainfall in Nigeria*. Lagos, 1928.

Thuillier, H. L., and Smyth, R. A. *A Manual of Surveying for India, Detailing the Mode of Operations on the Trigonometrical, Topographical and Revenue Surveys of India*. London, 1875.

"Timber and Forestry in Western Australia." *Indian Forester* 28, 1902.

"Timber in Nigeria." *Indian Forester* 33, 1907.

"Timber Needs of the Empire." *Journal of the Royal Empire Society* 20, 1929.

"Timber Resources of Liberia." *Indian Forester* 31, 1905.

Tristram, Henry Baker. *The Natural History of the Bible*. London, 1911.

Troup, Robert Scott. *Colonial Forest Administration*, Oxford, 1940.

Tserofski. "Village Forests." *Indian Forester* 18, 1892.

Tubbs, Colin R. *The New Forest: an Ecological History*. Newton Abbot, 1968.

Turner, Frederick Jackson. "The Significance of the Frontier in American History." *Annual Report of the American Historical Association for 1893*, Washington, DC, 1894.

Turrill, W. B. *Joseph Dalton Hooker: Botanist, Explorer and Administrator*. London, 1963.

Tuscan, C. P. "Forest Administration in the Central Provinces." *Indian Forester* 17, 1892.
  "Forest Administration in the Central Provinces." *Indian Forester* 19, 1893.
  "The Mythology of Forest Fires." *Indian Forester* 18, 1892.
United States, Department of State. *Forestry in Europe: Reports from the Consuls of the United States.* Washington, DC, 1887.
Unwin, Harold. "Canadian Forests and Forestry." *Indian Forester* 52, 1904.
  "The Prohibition of Grass Burning and its Effect on the Game of the Country." *Indian Forester* 31, 1905.
  West African Forests and Forestry. London, n.d.
Vagrant. "Located Fellings: a First Step Towards Regular Working Plans." *Indian Forester* 19, 1893.
Wadia, R. A. *The Bombay Dockyard and the Wadia Master Builders.* Bombay, 1955.
Walker, H. C. "Fire Protection in Teak Forests in Lower Burma." *Indian Forester* 28, 1902.
Walker, Ian Campbell. *Report of the Forest Department, Madras Presidency, for the Year 1885–86.* Madras, 1887.
  Reports on Forest Management in Germany, Austria, and Great Britain. London, 1873.
  Reports on Forest Management in the Madras Presidency. Madras, 1873.
Ward, H. G. *Hansard,* 3rd series, 33.
Watts, William A. "Europe." In *Vegetation History,* ed. B. Huntley and T. Webb III. Dordrecht, 1988.
Webb, W. *The Great Plains.* Boston, 1931.
Wells, H. G. *The Outline of History: Being a Plain History of Life and Mankind.* New York, 1921.
"West Indian Timbers." *Indian Forester* 28, 1902.
"Western Australian Timbers." *Indian Forester* 26, 1900.
White, Gilbert. *The Natural History of Selborne.* Boston, 1975.
White, Lynn, Jr. "The Historical Roots of our Ecological Crisis." In *Machina ex Deo: Essays in the Dynamism of Western Culture.* Cambridge, MA, 1968.
White, R. "American Environmental History: the Development of a New Historical Field." *Pacific Historical Review* 54, 1985.
Wicht, C. L. and Ackerman, D. *Our Green Heritage: a Book about Indigenous and Exotic Trees in South Africa, about Trees and Timber in our Cultural History and about our Extensive Silvicultural, Forestry and Timber Industries.* Cape Town, 1973.
Wild, A. E. *Report on the Forest in the South and West of the Island of Cyprus.* London, 1879.
Williams, Michael. *Americans and their Forests: a Historical Geography.* Cambridge, 1989.
  "The Relations of Environmental History and Historical Geography." *Journal of Historical Geography* 20, 1994.
Williamson, R. M. *Reports on the Forests and Lac Industry in Rewah State, Central India.* Allahabad, 1906.
Wilson, James Fox. "Water Supply in the Basin of the River Orange, or Gariep, South Africa." *JRGS* 35, 1865.
"Woods and Forests of the Soudan." *Indian Forester* 33, 1907.
"The Woods of Samoa." *Indian Forester* 22, 1896.
Worlsey, T. S., Jr. *Studies in French Forestry.* New York, 1920.
"Working of State Foresters in Russia." *Indian Forester* 29, 1903.

World Conservation Monitoring Centre. *1992 Protected Areas of the World: a Review of National Systems*. Cambridge, 1992.

Worster, Donald. *American Environmentalism: the Formative Period, 1860–1915*. London, 1973.

The Ends of Earth: Perspectives on Modern Environmental History. New York, 1988.

Nature's Economy: a History of Ecological Ideas. Cambridge, 1995.

Worthington, E. B. *Sciences in Africa*. Oxford, 1938.

Index

Cambridge Studies in Historical Geography

Titles marked with an asterisk* are available in paperback.